Elena Malofeeva

Meta-Analysis of Math Instruction with Young Children

D1783979

Elena Malofeeva

Meta-Analysis of Math Instruction with Young Children

What do we Really Know About How to Teach Mathematics to Preschoolers and Kindergartners?

VDM Verlag Dr. Müller

Impressum/Imprint (nur für Deutschland/ only for Germany)

Bibliografische Information der Deutschen Nationalbibliothek: Die Deutsche Nationalbibliothek verzeichnet diese Publikation in der Deutschen Nationalbibliografie; detaillierte bibliografische Daten sind im Internet über http://dnb.d-nb.de abrufbar.

Alle in diesem Buch genannten Marken und Produktnamen unterliegen warenzeichen-, marken- oder patentrechtlichem Schutz bzw. sind Warenzeichen oder eingetragene Warenzeichen der jeweiligen Inhaber. Die Wiedergabe von Marken, Produktnamen, Gebrauchsnamen, Handelsnamen, Warenbezeichnungen u.s.w. in diesem Werk berechtigt auch ohne besondere Kennzeichnung nicht zu der Annahme, dass solche Namen im Sinne der Warenzeichen- und Markenschutzgesetzgebung als frei zu betrachten wären und daher von jedermann benutzt werden dürften.

Coverbild: www.purestockx.com

Verlag: VDM Verlag Dr. Müller Aktiengesellschaft & Co. KG
Dudweiler Landstr. 99, 66123 Saarbrücken, Deutschland
Telefon +49 681 9100-698, Telefax +49 681 9100-988, Email: info@vdm-verlag.de
Zugl.: South Bend, University of Notre Dame, 2004

Herstellung in Deutschland:
Schaltungsdienst Lange o.H.G., Berlin
Books on Demand GmbH, Norderstedt
Reha GmbH, Saarbrücken
Amazon Distribution GmbH, Leipzig
ISBN: 978-3-639-19695-5

Imprint (only for USA, GB)

Bibliographic information published by the Deutsche Nationalbibliothek: The Deutsche Nationalbibliothek lists this publication in the Deutsche Nationalbibliografie; detailed bibliographic data are available in the Internet at http://dnb.d-nb.de .
Any brand names and product names mentioned in this book are subject to trademark, brand or patent protection and are trademarks or registered trademarks of their respective holders. The use of brand names, product names, common names, trade names, product descriptions etc. even without a particular marking in this works is in no way to be construed to mean that such names may be regarded as unrestricted in respect of trademark and brand protection legislation and could thus be used by anyone.

Cover image: www.purestockx.com

Publisher:
VDM Verlag Dr. Müller Aktiengesellschaft & Co. KG
Dudweiler Landstr. 99, 66123 Saarbrücken, Germany
Phone +49 681 9100-698, Fax +49 681 9100-988, Email: info@vdm-publishing.com

Printed in the U.S.A.
Printed in the U.K. by (see last page)
ISBN: 978-3-639-19695-5

CONTENTS

TABLES

CHAPTER 1

INTRODUCTION

Improving mathematical skills in American children is one of the top
priorities according to *America 2000: An Education Strategy*: By the year 2000, U.S.
students will be the first in the world in science and mathematics achievement
(Education, 1991). However, at the beginning of the twenty-first century research
studies still indicate that most children are behind in their development of
mathematics skills and as a result they perform lower on mathematical achievement
tests than students from other industrialized countries (Geary, 1995; Stevenson, Chen,
& Lee, 1993).

A number of training studies have looked at the relationship between various
mathematics skills and training in preschool and kindergarten children (Baroody,
1987b, 1989, 1992a, 1999; Chao, Stigler, & Woodward, 2000; Clements, 1984a;
Clements, Swaminathan, Hannibal, & Sarama, 1999; Dixon, Carnine, Lee, Wallin, &
Chard, 1998; Pasnak, 1987; Solter & Mayer, 1978; Sophian, 1998, 2002; Squire &
Bryant, 2002; Tzuriel, Kaniel, Kanner, & Haywood, 1999). Many researchers
proposed to reexamine mathematics instruction during preschool years in order to
improve on mathematics achievement, skill acquisition, and development of number

1

sense (e.g., Geary, 1994). A growing number of teachers and educational researchers suggest using a guided method of teaching children mathematics where a teacher is a coach helping children discover facts about quantities at their own pace (Brown & Campione, 1994; Carpenter et al., 1999; Kroesbergen & van Luit, 2002; Solter & Mayer, 1978). Yet, there is also evidence of efficacy of direct instruction (e.g., Phillips & Stipek, 1993; Stipek et al., 1998). Despite attention that is placed on teaching mathematics, very few quantitative analyses of mathematics training with younger children have been conducted in order to see what instructional techniques are successful with young children (Dixon et al., 1998). The study proposed here reflects an attempt to explore two main hypotheses: 1) How beneficial is early instruction in the acquisition of mathematical skills and knowledge (e.g., counting, addition, fractions)? and 2) What interventions are the most efficient in preschool and kindergarten years (e.g., Is discovery instruction more beneficial than direct instruction?) These two main hypotheses were investigated by means of a quantitative literature synthesis, a meta-analysis. Learning about good instructional practices might help teachers, psychologists, researchers and other professionals create teaching/learning experiences that positively affect children's performance in the area of mathematics.

The introduction consists of three chapters. Normative development of mathematics skills in young children is briefly described in the first chapter. Chapter 2 consists of 3 sections. The first section in Chapter 2 describes current trends in the area of mathematics instruction. In the second section we distinguish between two main types of instructional approaches: direct (i.e., teacher-centered) and guided (i.e., child-centered). We then proceed to the literature review on various instructional practices for most of the important mathematics skills developing during preschool and kindergarten years. The third chapter briefly explains how a meta-analysis is usually conducted. We then describe this meta-analytic study and main hypotheses of interest.

The goal of this research is to quantitatively summarize existing research findings in the area of mathematics instruction with three to six-year-old children and to identify instructional practices and approaches that yield high effect sizes. Since the basic foundation of mathematical knowledge is built during preschool years, the type and methods of instruction used during this critical period may be extremely important, especially if they affect children's later performance and mathematical achievement and further development of problem-solving abilities (Pellegrini & Stanic, 1993; Saxe, 1991).

Normative Development

Children have some knowledge and understanding of mathematics before they enter school. Such learning includes, as outlined in Principles and Standards for School Mathematics released by The National Council of Teachers of Mathematics (NCTM), a broad variety of skills (e.g., problem solving, estimation, geometry and spatial sense, measurement, fractions and decimals) (NCTM, 2000). We have come to know a number of different facts about how young children acquire and develop these skills. A number of researchers have proposed that all the skills and knowledge that children acquire in preschool and kindergarten lead to their development of number sense (Gersten & Chard, 1999). Number sense is a construct, analogous to phonemic awareness in reading research, that might help researchers put all the pieces of a puzzle on how young children learn mathematics together. According to Gersten and Chard: "Number sense ...refers to a child's fluidity and flexibility with numbers, the sense of what numbers mean, and an ability to perform mental mathematics and to look at the world and make comparisons" (p.19-20). The current literature review indicates that number sense develops in children before they enter school. Since number sense has already been linked to later achievement and lack of it to possible learning difficulties and school failure, it seems important to point out that three to six year-old children not only acquire basic mathematics skills described above but they also develop flexible understanding of mathematics. Instructional implications of this number sense development are discussed in Chapter 2.

During the preschool years children learn number words, recognize numerical symbols, and start using these symbols quantitatively. There is substantial research evidence that: (1) many young children know how to count (Arai, 1984; Baroody, 1992b; Baroody & White, 1983; Bertelli, Joanni, & Martlew, 1998; Clements, 1984a; Gallistel & Gelman, 1990; Newman & Berger, 1984; Wiegel, 1998); (2) many have some understanding of addition and subtraction (Baroody, 1987b, 1989; Baroody & Wilkins, 1999; Campbell & Maertens, 1988; Fuson, 1992; Fuson & Kwon, 1992; Geary, 1994; Groen & Resnick, 1977; Hughes, 1986; Secada, Fuson, & Hall, 1983; Siegler, 1987); (3) they can compare things around them (e.g., state that 8 is more than 2) (Kotovsky & Gentner, 1996; R. C. Williams, 1980); (4) they use strategies in problem-solving (Carpenter et al., 1999; Geary, 1994; Hughes, 1986; Swanson,

4

1992); (5) many can match and sort shapes and sizes (Clements et al., 1999; Tyler, Allen, & Pasnak, 1983); and (6) they start acquiring some measurement concepts (e.g., they begin to understand the concept of half and a whole and its parts) (Boisvert, Standing, & Moller, 1999; Fischer, 1990; Hunting & Sharpley, 1988; Kimchi, 1993; Sophian, Garyantes, & Chang, 1997).

In the past, Piaget emphasized the lack of children's general cognitive abilities to understand numerical knowledge. He believed that young children should not be given instruction early as they were able to discover facts and understand their meanings better on their own (Piaget, 1965). Through research studies we now know that young children possess many basic numerical skills (like counting or addition) and they can reason about numbers (e.g., understand that numbers have some quantity to them) but children experience difficulties (e.g., they can skip objects or unable to understand the reason behind counting) and they do not efficiently use strategies that older children and adults use (e.g., basing decisions on perceptual characteristics of shapes instead of functional) (Sophian, 1987, 1988; Sophian & McCorgray, 1994).

The normative development of mathematics skills described above is known to be affected by a number of factors: SES (Davie, Butler, & Goldstein, 1972; Ginsburg & Russell, 1981; Pellegrini & Stanic, 1993; Saxe, 1991); parents' involvement into teaching mathematics at home (Anderson, 1997; Ginsburg, Bempechat, & Chung, 1992; Leder, 1992; Starkey & Klein, 2000); characteristics of the task (Jordan, Huttenlocher, & Levine, 1992); language abilities (Gallistel & Gelman, 1992); and play (Yawkey, not known). For example, a one-year difference

5

was reported for in children's overall mathematical abilities between working and middle-class children (Davie et al., 1972). In other studies SES differences were reported for such specific skills as cardinality and more or less judgments (Ginsburg & Russell, 1981). All these factors need to be taken into consideration when teaching children mathematics.

Research on teaching the following mathematical skills is discussed in the next chapter: counting, understanding the numerical symbols, whole number computations (addition, subtraction), measurement (length, area, volume, quantity, weight, part-whole relations, time), comparisons and classifications, and geometry and spatial sense because these skills begin to develop before children attend school. For each skill, recent research findings on instructional aspects for young children are summarized.

CHAPTER 2

REVIEW OF RELEVANT LITERATURE

Emergent mathematics education research focuses on how young children learn mathematics and what environments/types of teaching strategies/interventions positively affect their learning. This section presents this research in three main parts. At first, the current view of mathematics instruction in young children as promoted in the National Council of Teachers of Mathematics (NCTM) and NAEYC standards, goals and position statements is offered. We then describe two main instructional approaches presently employed by educators: direct (teacher-centered) and guided (child-centered). Last, an overview of emergent mathematics education research is presented.

The Current View of Emergent Mathematics Instruction

The NCTM has recently released Principles and Standards for School Mathematics (2000) where some information on early mathematical development is provided for the first time. For example, young children should "develop understanding of … the ordinal and cardinal numbers" (p.78), "connect number words with quantities they represent" (p.79), "associate number words with small collections of objects" (p.79), or "solve addition and subtraction problems based on counting strategies" (NCTM, 2000). The NCTM also published a position statement, *Early Childhood Mathematics Education* in 1991. According to this statement, "developmentally appropriate childhood mathematics instruction should meet the needs of the individual learners at different stages of readiness by considering the influences of cultural backgrounds, prior experiences, learning styles, and cognitive abilities" (p.17). Developmentally appropriate practices (DAP) in the area of mathematics stress training that is individually and age appropriate. Such practices incorporate real-world contexts, children's ideas, and experiences using a variety of materials and supplies (Hart, Burts, & Charlesworth, 1997). In developmentally appropriate classrooms, for example, children might solve measurement problems using an equal-arm balance. A teacher brings two similar-looking bags into the classroom: one with heavy blocks, the other one with light blocks. Children think that both bags will weight the same. However, soon they realize that even if the bags look the same, they might have different types of blocks in them and therefore, their weights are different. A teacher allows children to create their own theories of the world around them and then test these theories against reality. Baroody (Baroody, 1987a) provides a description of a developmentally appropriate approach to teaching mathematics in the following way: "Teaching mathematics is essentially a process of translating mathematics into a form children can comprehend, providing experiences that enable children to discover relationships and construct meanings, and creating opportunities to develop and exercise mathematical reasoning and problem-solving abilities" (p.40). Therefore, "early childhood mathematics education, for young children aged 3-8, should be developmentally appropriate (NCTM, 1994-95, p.16).

Different researchers describe mathematics instruction using different terms. For example, Baroody (1998) views teaching as a process of teaching for skill mastery (a skills approach), teaching for understanding (a conceptual approach) or teaching for problem-solving (a problem-solving approach). Hiebert and Carpenter talk about two major ways of teaching mathematics: top-down and bottom-up (Hiebert & Carpenter, 1992). The goal of the former is to move students from a novice to an expert (it is very similar to Baroody's skills teaching approach). The latter approach emphasizes creating new knowledge based on what students already know without being shown directly the most effective way (more or less a combination of Baroody's problem-solving and conceptual approaches) (Ginsburg & Baron, 1993). Vygotsky (1978) in his theory differentiated between spontaneous and formal concepts. The bottom-up approach stresses the importance of development of both types of concepts and bases formal mathematical knowledge on spontaneous knowledge that children have developed by interacting with their environment during their preschool years. Consistent with Piagetian and Vygotskian ideas, bottom-up instruction is considered to constitute the current movement in the area of mathematics instruction with the focus on learning for understanding and using developmentally appropriate paradigm.

Two Main Approaches to Instruction

Several debates characterize the literature on young children's developing understanding of math concepts and skills. Researchers are interested in: a) what methods (i.e., naturalistic, informal or formal (structured)) are the most beneficial to the students and at what age; b) whether preschool instruction should include training in logico-mathematical reasoning (i.e., Piagetian concepts) as core components; and c) what is the relationship between acquiring mathematics skills (procedural knowledge) and understanding (conceptual knowledge). As part of the first debate, a vast majority of researchers recognizes guided (child-centered) or direct (teacher-centered) approaches to teaching mathematics. Below we describe each of the two main approaches to instruction and their implications for children's development.

Guided Instruction. Guided instruction is an approach where children discover relationships and constructs that organize the world around them and thereby construct their mathematical knowledge (Charlesworth & Lind, 1990; Cobb, 1994; Piaget, 1965; Resnick, 1989). For example, children can use blocks and learn more about different shapes and their properties. Or when picking chairs for all children in day care, children can compare the number of chairs with the number of children. Learning becomes meaningful and fun when used in the context of appropriate activities. In guided discovery learning, the attention is shifted to the child (or is child-centered) and the process of learning about quantities becomes a true dialogue between participants.

Some research favors the guided approach. Children display more positive academic, motivational, emotional and behavioral outcomes in child-initiated programs (Bryant, Clifford, & Peisner, 1994; Stipek, Feiler, Daniels, & Milburn, 1995). Research on developmentally appropriate practices indicates that children benefit from child-centered instruction (Bryant, Clifford, & Peisner, 1991; Stipek et al., 1995). Both short-term and long-term (children were followed in elementary grades) positive effects were established (Burts, 1993). In an interesting study by Solter and Mayer, 43 preschoolers learned the concept of one-to-one correspondence by matched discovery, expository (i.e., direct instruction), or observation methods (i.e., control condition) of instruction. Performance on a subsequent posttest revealed no difference between groups on short-term recall and near transfer, but the discovery group excelled on far transfer (conservation) and delayed recall (Solter & Mayer, 1978). Authors reiterated the advantage of teaching young children using a non-direct approach and pointed to the importance of using measures of near and far transfer for better identification of successful teaching strategies.

Direct Instruction. Direct instruction is an instructional approach where a teacher explicitly instructs students on learning strategies by modeling and explaining why, when, and how to use them. Review of the literature describes direct instruction as having the following characteristics: 1) breaking a task into smaller units; 2) administering feedback; 3) providing a pictorial and diagram representation; 4) teacher modeling a skill; 5) a teacher providing set materials at a rapid pace (Swanson, Carson, & Saches Lee, 1996). Often, a child has to answer specific questions that are chosen by an adult and are oriented towards fulfilling a specific educational goal.

Those researchers who postulate the importance of a more direct approach in emergent mathematics instruction emphasize the active role a teacher should play in the classroom and the importance of feedback about children's performance in developing children's motivation (Gifford, 1995). Swanson and colleagues, for example, analyzed 78 intervention studies conducted with children and adolescents with learning disabilities (different domains were analyzed, not just mathematics) (Swanson et al., 1996). This group of researchers found that intervention studies that produced the highest effect sizes were related to derivatives of cognitive and/or direct instruction. This finding is consistent with their later meta-analysis of 180 intervention studies where they also found that the direct instruction was beneficial (Swanson & Hoskyn, 1998).

No one model can describe how to best teach children mathematics. As a result, early childhood education should not be seen as an educational issue of one size fits all. The answer to the question of what type of intervention style (e.g., direct versus guided) is truly developmentally appropriate depends on a variety of factors (e.g., culture, background experience, the goal of teaching). Some children need more structure than others. Children from some linguistic and cultural backgrounds may be more accustomed to direct instruction; therefore, guided learning might create an atmosphere of tension and embarrassment for them. Learning how various child, teacher, environment, and cultural characteristics mediate learning through various interventional approaches will help us find better fits for particular children with particular needs. Sophian (Sophian, 2003) also suggests adopting a prospective developmental perspective to teaching mathematics with a careful assessment of what children need to learn early on in order to be conceptually and procedurally ready to acquire mathematical concepts later in school.

Detailed next is the literature review on effective instructional techniques and intervention programs that are known to contribute to children's learning in the area of mathematics. After reporting on results of major meta-analyses in the area of mathematics intervention, we describe research findings on different characteristics of instruction by specific mathematical skill.

Literature Review on Instructional Practices

Our examination of intervention studies with preschool and kindergarten children suggests a number of distinct features that characterize present day instructional research in early mathematics: establishing instructional classifications, synthesizing existing primary intervention studies in mathematics using a meta-analysis, and further investigation of intervention strategies at the level of primary studies.

Instructional Classifications. A number of instructional classification schemes have been proposed by researchers. We present four such classifications: Dixon et. al (1998), by Swanson and Hoskyn (1998), Maccini and Hughes (1997), and Lee (2000) (see Table 1). These instructional classifications are very useful in a meta-analytic study as well as when designing successful intervention strategies at the level of primary research studies.

TABLE 1

INSTRUCTIONAL CLASSIFICATIONS

Dixon et. al. (1998)	Swanson and Hoskyn (1998)	Maccini and Hughes (1997)	Lee (2000)
Use manipulatives	Sequencing	Behavioral (reinforcement, modeling, demonstration, feedback, and other teacher-directed)	Using strategy-based intervention
Work among peers	Drill repetition and practice review	Cognitive (goal-setting, self-instruction, self-monitoring, or word-problem solving strategies)	Sequencing examples
Design of instruction	Anticipatory or preparation response	Alternative Delivery Systems (computer assisted or video-disc instruction, and cooperative learning)	Adjusting difficulty levels
Computers and technology	Structured verbal teacher-student interaction		Using alternative representation mode
Grouping	Individualization and small group		Identifying and explicitly teaching prerequisite skills
Reinforcement or motivation systems	Novelty		Using peer tutoring between two students
Idiosyncratic studies	Strategy modeling and attribution training		Working within a group of peers
	Probing-reinforcement		Using teacher modeling
	Non-teacher instruction		Using guided practice
	Segmentation		Using motivation-based intervention
	Advanced organizers		Using mastery criterion
	Directed response/questioning		Using specific feedback
	One-to-one instruction		Individualizing instruction by skill analysis
	Control difficulty or processing demands of task		Using technology as delivery media
	Technology		Using technology for independent practice
	Elaboration		
	Modeling by teacher of steps		
	Group instruction		
	Supplement to teacher involvement other than peers		
	Strategy cues		

14

Swanson and colleagues analyzed 180 intervention studies with children with learning disabilities using their intervention classification as specified in Table 1, Column 2 (Swanson & Hoskyn, 1998). They reported an overall mean effect size for instructional intervention of 0.79. Effect sizes were higher for those models that included a combination of direct and strategy instruction, that controlled task difficulty, included small interactive groups, directed responses, and questioning of students.

Dixon and colleagues developed their own classification of interventions in mathematics. They reviewed high quality intervention studies using criteria mentioned in Table 1. They identified those studies with high internal/external validity and significant improvements in performance on various mathematics measures (Dixon et al., 1998). Simple vote counting, however, confounds treatment effect and sample size and therefore, might be misleading.

Intervention Meta-Analyses. Recently, due to an increased number of primary research studies that promoted mathematics training, a number of meta-analyses on intervention in the area of mathematics were published (Baker, Gersten, & Lee, 2002; Butler, Miller, Lee, & Pierce, 2001; Kroesbergen & Van Luit, 2003; Mastropieri, Bakken, & Scruggs, 1991; Swanson, 2000; Swanson et al., 1996; Swanson & Hoskyn, 1998). Mastropieri and colleagues, for example, identified 25 mathematics intervention studies that used a total of 519 mentally retarded individuals (aged 6.33-.53 years) (Mastropieri et al., 1991). Three areas: (a) basic skills and concepts, (b) rule learning and problem-solving, and (c) applications such as use of time, money, and measurement skills were identified in their meta-analysis. Overall, intervention studies reported positive effect sizes for training in mathematics. Swanson and Hoskyn reported that cognitive strategy and direct instruction models were effective tools in remediating the academic difficulties for children with learning disabilities (Swanson & Hoskyn, 1998). They also reported that interventions that varied from control conditions in terms of setting, teacher, and number of instructional steps yielded larger effect sizes than studies that failed to control for such variations. In her other meta-analysis with students with learning disabilities, Swanson found that intervention math studies had an overall effect size of .40 (Swanson, 2000).

Baker and colleagues identified 15 studies with low-achieving students (Baker et al., 2002). These authors reported the following instructional characteristics led to improvements in the mathematics achievement of students experiencing difficulties in mathematics: (1) providing teachers and students with data on student performance; (2) using peers as tutors or instructional guides; (3) providing clear, specific feedback to parents on their children's mathematics success; and (4) using principles of explicit instruction in teaching math concepts and procedures.

Miller and colleagues also identified effective instructional practices for students with mental retardation (Butler et al., 2001). They reported that such techniques as constant-time delay, peer tutoring, time trials, and direct instruction proved beneficial in improving mathematics skills. Further, students with mental retardation were successful in learning cognitive strategies. The authors concluded that more studies need to be conducted in secondary schools and in inclusive settings.

In Kroesbergen and Van Luit's review of 58 studies of mathematics interventions for elementary special needs students, direct instruction and self-instruction were found to be more effective than mediated instruction (Kroesbergen & Van Luit, 2003). They found lower effect size for peer tutoring (.87 versus .96 when tutoring was not used). Other studies looked at math interventions and problem solving skills (Jitendra et al., 1998) or children with learning disabilities (Swanson & Hoskyn, 1998; Swanson, Hoskyn, & Lee, 1999). Most of these studies, however, are based on samples of older children. They did not include primary studies with preschool children.

    Primary Intervention Studies. In addition to existing meta-analysis and proposed classifications of intervention characteristics, a major trend in early mathematics education research is to understand the effectiveness of *specific* teaching methods used to teach mathematical concepts to specific population of students. Such methods as the use of fingers or gestures in counting (VanDevender, 1986); ways of recording different number of objects (e.g., tallying) (Hughes, 1996); types of tasks (e.g., using concrete/hypothesized objects or solving a problem abstractly) (Hughes, 1981); use of number books (Hong, 1996; Jennings, Jennings, Richey, & Dixon-Krauss, 1992) were all explored. Unanimous empirical support exists on using manipulatives, hands on activities and real objects in various mathematical activities (Kamii, Lewis, & Kirkland, 2001; VanDevender, 1986). Mixed support exists on effectiveness of number books. Hong, however, reported improvement in classification, number combinations, and shape tasks as a result of teaching mathematics through children's literature (Hong, 1996). Others looked at the effectiveness of special programs (e.g., Bright Start Program) created and implemented in the preschool or kindergarten samples (Bobis & Whitton, 1999; Southard & May, 1996; Tzuriel et al., 1999).

Across theories and teaching styles, research findings indicate that excellent mathematics instruction can be characterized by the following characteristics: a teacher uses situations that are personally meaningful to the children, provides opportunities to make problem solving choices (instead of telling them the right answer), provides peer interactions where exchange of ideas becomes possible and knowledge and skills are further practiced and constructed (i.e., cooperative learning groups), develops internal motivation for learning, uses variety of materials and situations (e.g., a sand-box, construction paper, manipulatives), allows for practice and transfer of acquired knowledge and skills, uses various assessment methods (i.e., static and dynamic), and uses new technologies (e.g., computers, calculators) (Baroody, 1987a; Clements & Sarama, 2003; Fennema, Carpenter, & Lamon, 1991; Geary, 1994; Ginsburg & Baron, 1993; S. Griffin, Case, & Capodilupo, 1995; C. K. Williams & Kamii, 1986).

We will now turn to the specific description of successful intervention strategies and techniques with three-six year old children. We will describe the recent research findings by specific skill.

Instruction on Number Sense

Number sense is child's potential to be flexible with numbers, especially, in new mathematical contexts, and child's intuitive feel for numbers and their quantities. It has been operationally defined as the ability to make quantitative judgments and comparisons, to understand the characteristics of the number system and relative

magnitude of numbers, to recognize faulty mathematical computations, to calculate

operations mentally, to understand quantities and their properties, cardinality,

ordinality, and to develop referents for measures of common objects and situations in

their environment (Baroody & Wilkins, 1999; Gersten & Chard, 1999; NCTM, 1994-

95; Pike & Forrester, 1997). Number sense is a construct, analogous to phonemic

awareness in reading research, that emphasizes conceptual understanding of

numeracy while relying heavily on computational mastery (Gersten & Chard, 1999)

Research on emergent mathematics has promoted the inclusion of number

sense instruction in the curriculum for young children. Part of that effort includes

identifying strategies that young children use and teaching efficient strategy use even

before children enter elementary school. Siegler (1988), for example, indicated that

children learn to start counting from the larger number when adding (e.g., starting to

count with five, six, seven, eight in "3+5"). However, some children (especially, low

SES and students with learning disability) fail to learn this strategy. Therefore, these

children might benefit from strategy instruction. Shrager & Sigler (1998) in a recent

study warn educators that strategy instruction is slow in young children and it might

require some effort and time. Case (1998) and Griffin (Gersten & Chard, 1999)

suggest using three (instead of multiple modalities) representational systems in

kindergarten classes (mathematical symbols (e.g., addition sign), a vertical

thermometer, and a system similar to Candy land game) as well as verbalizing taught

strategies as specific ways of developing number sense. Such early interventions

19

suggested by the researchers help children gain experiences that they might not always gain in preschool and at home.

In a number of studies by Case and colleagues (Case & Griffin, 1990; S. Griffin et al., 1995; S. Griffin, Case, & Siegler, 1994), children were introduced to "a mental number line" that was used to develop the central conceptual schema needed to understand the environment in a quantitative fashion. Teachers concentrated on certain properties of numbers (e.g., moving up one place on a number line means increasing quantity by one), asked interesting questions about numbers, let students interact during the process of finding solutions to problems involving numbers, and stimulated and promoted students' mathematical reasoning (Geary, 1995; S. Griffin et al., 1994). Chao and colleagues (Chao et al., 2000) also developed a number of interesting games that stimulated intuitive sense of number, even though the goal of their study was not related to increase in children's math performance.

To date, most researchers have been interested in understanding how number sense develops in older school children; very few have considered addressing instructional aspects of number sense in young children. In Malofeeva et. al. (Malofeeva, Day, Saco, Young, & Ciancio, unpub.) study, researchers assessed seven different skills (i.e., counting, number identification, comparison, addition and subtraction, ordinality, and number-object correspondence), each of which has been included in the definitions of number sense. In addition, the question of how instruction on two basic number sense skills (i.e., counting and number identification) affected children's performance on other number sense skills was explored. Forty 3- to 5-year-old children attending Head Start took pre- and posttests on the Number

Sense Test. After the pretest, children participated in 6 instructional sessions learning either about number sense-related skills (i.e., instructional condition) or about insects (i.e., attention control). Understanding of number as assessed by performance on the counting and number identification scales of the Number Sense Test improved significantly from pre- to posttest for children in the instructional condition. Except for addition and subtraction, training effects did not generalize to other number-sense skills (i.e., ordinality, comparison, and number-object correspondence).

Instruction on Counting

During preschool years children learn to count (Baroody, 1987b; Bertelli et al., 1998; Clements, 1984b; Fuson, 1995; Fuson, Lyons, Pergament, Hall, & et al., 1988; Munn & Stephen, 1993; Pepper & Hunting, 1998). Counting is often considered a foundational skill that influences other math skills. Therefore, teaching children counting is often discussed in relationship with such skills as number sense, geometry, comparison, addition and subtraction, and fractions.

Teaching number skills in preschool years has received unanimous support (Carpenter et al., 1999; Geary, 1994; Gelman & Gallistel, 1978; Young Loveridge, 1989). Educators can provide various opportunities for children to practice their counting skills by using manipulatives and real objects. In Groen and Resnick's study , for example, children were able to use mental counting instead of concrete counting when adding only after 12 to 20 weeks of instruction. In Baroody's study (1987b), however, even after 6 months of instruction (twice a month for 20-30 minutes) some children still continued to count with objects only. There is some evidence that direct instruction in counting strategies is beneficial (Geary, 1994).

Wiegel (1998) investigated how counting strategies were affected by cooperative learning (Wiegel, 1998). In his experiment kindergarten students were taught weekly for 4 months. Four different strategies evolved as a result of collaboration: counting objects side by side, counting all objects at the same time, taking turns, and counting together toward a common goal. There appeared to be a positive correlation between the level of counting skills and an ability to work cooperatively.

In the past, research training on counting has been linked to improvement on such tasks as addition and subtraction, some evidence also exists that knowledge about counting seems to develop independently from understanding of addition and subtraction (Baroody, 1987b; Baroody & Wilkins, 1999; Fuson, 1995). Training on counting was also linked to improvement on such tasks as classification and seriation (Clements, 1984b), addition and subtraction (Baroody, 1987b). However, presently logical operations are not considered to be prerequisites to learning how to count (Fuson, Secada, & Hall, 1983). Some evidence also exists that knowledge about counting seems to develop independently from addition and subtraction (Bertelli et al., 1998; Levine, Jordan, & Huttenlocher, 1992).

Instruction on Numerical Symbols

The research studies that investigate young children's identification of numerical symbols describe how they represent quantities. Efficacy of instructional aspects of numerical symbol development is limited. Hughes developed a training program where twenty four-year-old children played the Tins game (i.e., matching numerals to the number of bricks in different containers) and used magnetic numerals for eight sessions over a period of six weeks (Hughes, 1983). Most children in the study were able to match the numerals and understand that "+" meant adding some bricks and "-" meant taking some bricks away. However, transfer of knowledge did not occur: children were not able to use the acquired knowledge outside the game context.

Since there is a discrepancy between how young children represent number (i.e., pictographically or iconically) and symbolic representations of numbers, learning how to represent number on a piece of paper is very difficult for young children. One recommendation is that children should not be systematically taught written symbolic representation until elementary grades (Hughes, 1986).

Instruction on Addition and Subtraction

Contradictory to Piaget's claims, young children have some knowledge of addition and subtraction before they enter school (Baroody, 1987a, 1987b, 1989, 1992b, 1999; Baroody & Wilkins, 1999; Campbell & Maertens, 1988; Fuson, 1992; Fuson & Kwon, 1992; Geary, 1994; Groen & Resnick, 1977; Hughes, 1986; Secada et al., 1983). NCTM considers teaching addition and subtraction to preschool and kindergarten children developmentally inappropriate even though it is now known that children can further improve their existing skills through training (Groen & Resnick, 1977; Hughes, 1986).

23

Most children add or subtract using a variety of strategies: counting-on strategy (i.e., counting objects), using fingers and visual image of the objects (Hughes, 1986). In Groen and Resnick study (1977), for example, five-year-old children were taught to count two blocks, then three blocks, then the combined set. Despite the taught strategy, some children used shortened counting (i.e., counted "three, four, five" instead of starting counting blocks from one).

Instruction on Measurement and Fractions

During the preschool years children learn the concepts of surface and area (covers as much as, covers more, less, or least). They can use equal arm balance to estimate object weights. Many children can measure objects with the help of other objects (e.g., the car's length is the six little cars length) as well as explore how much space their parts of the body cover (e.g., tracing their hands and comparing them to their parents' tracings). Preschool children have some basic understanding of the requirements of measurement tasks and, therefore, they are able to develop some basic strategies to assess relative quantity, volume, length, etc. These strategies are, however, different from those used by adults (Geary, 1994; Miller, 1984). For example, when three-year-olds were asked to divide a hot dog among three people, they cut it into three pieces and gave each person a piece. However, pieces were of different length. Thus, researchers inferred that children's understanding of measurement was still prequantitative in nature (Miller, 1984).

Whereas Piaget focused his attention on what children cannot yet do in preschool years in the area of measurement, recent research studies investigate measurement skills that preschool children do have. Since such concepts of measurement as length, area, volume, weight, quantity were linked to children's ability to conserve, many research studies looked to determine if conservation skills could be improved through training in preschool and kindergarten children. There is evidence that even five year-old children are able to correctly conserve if they are trained to attend to correct dimensions (Clements, 1984b; Pasnak, Madden, Malabonga, Holt, & et al., 1996; Pasnak, McCutcheon, Holt, & Campbell, 1991). That is, even though children have difficulties understanding conservation, they nevertheless acquire important measurement concepts that lay a foundation for the later measurement experiences.

An interesting measurement perspective was proposed by Sophian who taught enumeration, measurement, and identification of relations among geometric figures using this paradigm (Sophian, 2003). In her study, 123 children were assigned to either an experimental mathematics condition (n=46), to a literacy control (n=48), or no-intervention control groups (n=29). Children in the experimental condition learned about the concept of a unit by counting and measuring the same quantity using different units, comparing different quantities, and discovering part-whole relationships. While Sophian's intervention included a variety of skills taught by trained Head Start teachers and parents throughout the school year, a foundation of the intervention was based on introducing an idea that units needed to be equal and that the result of counting was a function of the type of unit used. Children who received such measurement-based intervention outperformed children in the two control conditions on the mathematics subscale of the Developing Skills Checklist

(Sophian, 2003). The effect sizes were significant but not large and neither the fidelity of parents' instruction nor their instructional time was controlled for. While teaching children to conceptualize relationships between quantities based on the notion of the unit of measurement (as opposed to counting as the basis of further mathematics learning) is certainly an interesting avenue, more research is needed to further investigate instructional benefits and training maintenance and transfer.

Another curriculum was proposed by Fischer who trained 42 kindergartners on number concepts (Fischer, 1990). The children in the experimental group learned about cardinality by exploring parts into which sets of objects could be separated, counting sets and parts of objects, choosing numerals representing each picture set, and comparing parts among each other. The control group manipulation was controlled in terms of time, amount of manipulatives, worksheets, counting, and writing experiences. At the posttest, children who were taught through emphasis on part-whole relationship were better able to add and subtract as well as had a better understanding pf place value. Such training is recommended not only for the development of understanding of fractions but as a way to further foster addition, subtraction, and place value.

Instruction on Geometry and Spatial Sense

Children in preschool years are able to identify and match basic shapes. Most studies report that young children's geometric thinking is visual. That is, they are capable of recognizing components and simple properties of familiar shapes (Clements et al., 1999; Geary, 1994). Young children also learn such terms as next to,

26

near, toward, away, etc. that help them gain sense of the relationship of shapes with each other in space.

During preschool years, children also start to understand spatial relationships between objects. Teachers and parents can encourage children to develop their spatial senses when they use objects by inquiring and talking about what children play with: "Are your cars inside garage or outside? What is next to the red car?" Children like to play with blocks, and construction toys. At some point two-dimensional (e.g., puzzles) and one-dimensional shapes (e.g., alphabet letter parts) can be introduced. Training children identify shapes and match shapes has some positive outcomes (Sophian, 2003; White & Alexander, 1986). White and Alexander (1986), for example, trained four-year-old children to solve geometric analogy problems. Trained participants outperformed the control group and the results of the training were maintained for a one month period. Hong (1996) reported improvement in shape tasks (making triangles and squares by connecting dots) when teaching mathematics to fifty-seven kindergartners through using children's books.

Clements and Sarama developed an extensive curriculum called Building Blocks where children learned about spatial and geometric knowledge (as well as about numeric and quantitative knowledge) based on a variety of developmentally appropriate activities built on their interests and experiences (Clements & Sarama, 2003; Sarama & Clements, 2002). Computers, manipulatives, and print were three types of media that were used during such mathematical learning. Authors reported that effect sizes comparing children's performance in two conditions were .85 for number and 1.44 for geometry. Results of this study indicated strong positive effects of the types of materials used in Building Blocks project. Overall, children should be given multiple opportunities to learn about such shapes as circles, triangles, squares, rectangles, pentagons, and octagons both in preschool and at home using such media types as computers, manipulatives, and print.

Instruction on Comparison

Children in preschool year start making comparisons based on various dimensions. Training seems to improve children's performance on comparison tasks. For example, Holland and Palermo (1975) reported that kindergarten children were quite capable of learning more or less distinctions if properly trained. However, performance on conservation tasks did not improve as a result of such training. In Malofeeva and Day study (unpub.) three to five year-old children in both experimental conditions (activities and reading, activities only) significantly outperformed children in the control condition when making size, color, and function comparisons. However, their knowledge was not transferred to another familiar content domain (when comparing people's appearances). There was limited support for developing comparison skills when children were taught similarities and differences though comparing book characters and pictures while reading books to children (Malofeeva & Day, unpub.).

Instruction in the Piagetian Tradition

In his theory of child developm ent, Piaget emphasizes cons tructivist view of how m athematics should be taught. He stre sses the im portance of exploration and independence in child's m athematical developm ent and ineffectiveness of teaching before children are read y to learn (i.e., befo re they reach concrete operations s tage). Therefore, according to Piaget children need to be able to conserve first before being involved in any type of instruc tional practices. For example, in *Comments on Mathematical Education* (1973), Piaget argues that chil d's learning is a result of her general cognitive maturity and that a teach er cannot just pass his/her knowledge to a child because this knowledge needs to be reinvented by a child in her discovery of the world around her. Even though Piaget never specified how exactly teaching of young children should be carried out, his ideas po int to the im portance of us ing re al-life objects in understanding mathematics knowledge early on. For example, playing with water m ight help children see for them selves how conservation of volum e occurs. While learning is centered on children and thei r reconstruction of their environm ents, teacher's role is to provide these environments for them to explore.

Since Piaget developed his theory of logical-mathematical thinking, researchers have been trying to understand the importance of these concepts in preschool years and the necessity of emphasizing teaching such concepts as matching, sorting, seriation, and classification to three-six year-old children (Ciancio, Rojas, McMahon, & Pasnak, 2001; Ciancio, Sadovsky, Malabonga, Trueblood, & Pasnak, 1999; Gelman, 1982; Pasnak, Brown, Kurkjian, Mattran, & et al., 1987; Pasnak, Hansbarger, Dodson, Hart, & Blaha, 1996; Pasnak et al., 1991; Pasnak, Whitten, Perry, Waiss, & et al., 1995). According to Piaget, seriation, classification, one-to-one correspondence, and the ability to conserve under some perceptual transformations are presumed to underlie the development of number sense. Young-Loveridge (1987), for example, discusses young children's readiness to mathematics instruction and points that children are not developmentally ready to learn mathematics until they better develop more general logico-mathematical abilities. In one of his studies, Pasnak (1996) trained a group of kindergartners on unidimensional classification, seriation, and conservation. Children in the control condition received everyday mathematics instruction. Five months after the end of the study, children did not have any changes on a mathematics subscale of O-LSAT and a mathematics subscale of SESAT. Yet in their other study, Pasnak and colleagues (Pasnak & Campbell, 1991) successfully taught kindergarten children number conservation, classification, and seriation. Clements (1984b) also successfully trained four-year-old children on classification, seriation and counting.

There is growing concern that these concepts are mostly an extension of Piaget's theory and they are not based on the actual way children think or learn mathematics (Gifford, 1995). When Mohanty and Mishra trained four and five-year-old children for 4 weeks in either logical operations or number skills, the number skills group significantly outperformed the logical operations group on the number knowledge test, and the logical operations group significantly outperformed the number skills group on logical operations test (Mohanty & Mishra, 1994). Due to greater number skills transfer than due to logical foundations training, the authors claimed that as long as children received instruction on counting, explicit training in logical operations might not even be needed (Mohanty & Mishra, 1994). Thus, there is a continuing debate as to what the starting point of mathematics instruction might be: is it the development of logical thinking that might lead to number reasoning or should number understanding be developed first (Gifford, 1995; Pasnak et al., 1991).

Summary

In the eighties teaching/learning mathematics (partly in preschool, more so in kindergarten) was attempted mostly through direct instruction. A decade later mathematics instruction has become less teacher-directed and more child-centered and developmentally appropriate (Hart et al., 1997; Mathematics, 2000; NCTM, 2003). A constructivist approach to mathematics instruction together with Vygotskian sociocultural ideas have provided theoretical foundations to the present instructional practices in the area of emergent mathematics. In addition, Sophian (Sophian, 2003) suggests adopting a prospective developmental perspective to teaching mathematics with a careful assessment of what children need to learn early on in order to be conceptually and procedurally ready to acquire mathematical concepts later in school. Less and less emphasis is presently being placed on Piaget's theory and more on how to enhance specific mathematics skills and knowledge. Across different skills and instructional approaches, research findings point to one general conclusion: children develop their mathematical skills and knowledge long before they enter formal schooling. Training in mathematics seem to further enhance this development yet the magnitude of such learning averaged across studies and methods is not yet known.

Young children seem to be capable of learning a variety of skills including (but not limited to) counting, addition, subtraction, and measurement. The importance of including number sense activities in the instructional practices seems

31

logical. Most preschool programs emphasize counting small numbers, comparing, reciting number rhymes, reading number books, and developing problem-solving skills through games and play. According to Gifford (1995), most recommendations in the area of emergent mathematics instruction still include sorting, matching, and ordering. There have also been attempts to provide remedial instruction early on even before children start attending school (Van Luit & Schopman, 2000). Based on the literature review, we conclude that a growing number of educators and parents recognize the need to teach children mathematics early. There is a growing trend to develop long-term curricula (e.g., Building Block, Bright Start, Count Me In Too) where a combination of skills and approaches are used instead of one isolated specific skill or technique.

In the last decade a growing number of reviews and research syntheses have been published on mathematics interventions. For example, only in the last 2 years at least 2 meta-analyses have been conducted in the area of mathematics with older children. One study looked at what instructional practices contributed to improved mathematics achievement in children at risk for failure (Baker et al., 2002). Kroesbergen and Van Luit analyzed 58 studies of mathematics interventions for elementary special needs students (Kroesbergen & Van Luit, 2003). Both of these studies did not include studies with preschool children.

Other studies looked at math interventions and problem solving skills (Jitendra et al., 1998) or children with learning disabilities (Swanson & Hoskyn, 1998; Swanson et al., 1999). There have also been at least 2 meta-analyses with younger (i.e., preschool) children but in the area of literacy. One study focused on

32

phonological awareness in training studies (Bus & van Ijzendoorn, 1999). The other focused on benefits of joint book reading with preschoolers (Bus, van Ijzendoorn, & Pellegrini, 1995). Presently, no meta-analytic studies in mathematics training studies with preschoolers or kindergarteners have been conducted even though some qualitative literature reviews of main findings in this area are available (Charlesworth & Lind, 1990; Dixon et al., 1998). These qualitative reviews reveal a possible positive relationship between instruction and development of mathematics skills. However, simple vote counting might be quite deceiving as it confounds treatment effect and sample size. Studies with small sample sizes might not be significant because of low statistical power (Lipsey & Wilson, 2000). Therefore, at this point it is very hard to comment on the relationship between training in mathematics and children's performance on mathematics-related measures of achievement and specific measures of math skills with younger children.

CHAPTER 3

META-ANALYSIS

We begin this chapter by reviewing a meta-analytic method of research synthesis. We then describe the proposed study and identify main hypotheses.

Meta-Analysis

A single study is never perfect or definitive. Many similar studies with replicated results are needed to establish a scientific fact. Meta-analysis is an intriguing approach to the synthesis of research studies that has gained its popularity in the last ten years. In short, an effect size is calculated for each study. These standardized effects are then averaged across a sample of studies to get an index of an overall effect. Meta-analysis provides guidance and summary to researchers, its findings are usually easier to interpret. Meta-analysis is more objective than a traditional qualitative review because it is based on a set of primary studies and it uses statistical techniques (as opposed to simple vote counting) to account for variability in treatment effects. It often allows testing hypotheses that have not been tested in individual studies.

Meta-analysis also has some drawbacks. Many have claimed that meta-analyses compare apples and oranges. Critics point to the fact that while replications of the same methodology might be quite acceptable, combining findings from different studies might be problematic. Glass et al. (1981) argue that such comparison is what quantitative reviewers do and should do, but they should code apples as apples and oranges as oranges; they do not just throw all the fruit in one basket. It can lead to very different conclusions when available studies are few in number and/or if they are very heterogeneous. There are also numerous problems to consider: difficulty finding primary studies, different methodological quality of primary studies, publication bias-file drawer problem, inadequate statistics and procedure reporting, and multiple studies within a paper.

Meta-analysis involves the following steps: a) formulating the research questions/hypotheses; b) searching the literature trying to identify primary relevant studies; c) developing coding criteria; d) coding the primary studies; e) calculating effect sizes; f) analyzing statistically the effect sizes (e.g., multiple regression, the homogeneity test, etc.), and g) understanding the findings, conclusions, critical review of the studies, and future directions (Cook et al., 1992; Hunter & Schmidt, 1989; Lipsey & Wilson, 2000; Wachter & Straf, 1990).

The most important step in meta-analysis is formulating the research questions. Most hypotheses take into account previous theoretical/empirical/philosophical aspects in a particular research area where a number of primary research studies have been conducted to address similar hypotheses. Then an extensive literature search is conducted trying to locate unpublished studies and dissertations. Computer searches, manual searches, reference list searchers together with contacting researchers in a particular field are usually used to identify relevant studies.

Coding criteria vary from study to study. I have reviewed 10 meta-analytic studies in the fields of educational/developmental psychology that have been published in the last decade (Baker et al., 2002; Bus & van Ijzendoorn, 1999; Elbaum, Vaughn, Hughes, Moody, & Schumm, 2000; Fuchs, Fuchs, Mathes, & Lipsey, 2000; Gersten & Baker, 2000; Gersten, Schiller, & Vaughn, 2000; Marquis et al., 2000; Okolo, Cavalier, Ferretti, & MacArthur, 2000; Swanson, 2000; Swanson & Lussier, 2001). Most of the studies we reviewed included the following criteria: year of publication, gender, age, measures, sample size, published versus unpublished, design, type of statistic used to calculate the effect size, effect size, type of control group, type of program/therapy. The average number of criteria coded that was reported in those meta-analyses was 6 to 10. We used the coding criteria described above as well as additional criteria (each study was coded according to 191 different criteria) (see Appendix B for details).

In a meta-analysis, an effect size is computed for each individual study to measure the magnitude of the effect. Effect size can be conceptualized as a standardized difference. In the simplest form, effect size, which is denoted by the symbol "d", is the mean difference between groups in standard score form (i.e. the ratio of the difference between the means to the standard deviation). Cohen (1977) proposed the following values of effect sizes:

$$\text{small} \qquad\qquad d = .20$$

| medium | $d = .50$ |
| large | $d = .80$ |

The magnitude of effect sizes depends on the subject matter. After effect sizes
for each primary study are calculated, their variability is explored as a function of
various characteristics (e.g., year of publication, design, etc.). At the final step finings
of meta-analysis are explained in the light of possible limitations and future
directions.

Statement of the Problem and Hypotheses

Effective instructional techniques were identified through at least 7 meta-
analyses on mathematics interventions all conducted with older children (Baker et al.,
2002; Butler et al., 2001; Kroesbergen & Van Luit, 2003; Mastropieri et al., 1991;
Swanson, 2000; Swanson et al., 1996; Swanson & Hoskyn, 1998). The focus in this
meta-analysis was on analyses of available effective modes of mathematics
instruction with preschoolers and kindergarteners. We exclusively focused on this age
group while not restricting ourselves to a specific type of population served (e.g.,
children with learning disabilities or low achieving students).

While there have been a number of studies on mathematics training with
young children, questions about early mathematical intervention efficiency have been
raised due to a different degree of quality and validity of these primary studies. A
meta-analytic review could aid in clarification of the role of early interventions in
children's cognitive development and could influence future instructional practices
and standards.

37

The main question was to assess the impact of training on children's development of mathematics skills. Is such training effective? Does it work? Based on the qualitative review of the literature we expected to find a positive intervention effect on the development of young children's mathematics skills. We focused on the comparison of the children's performance in intervention programs with those children in the control groups. We used standardized effect size as a measure of this relationship due to the fact that measures were not similar across studies. The measures from which effect sizes were calculated were categorized as skill-specific or standardized.

The second question was to explore variables that had an impact on training. What influences the effectiveness of these interventions? Specifically, we were interested whether interventions that are characterized as direct, guided or mixed could contribute to significant differences in effect sizes. Literature review points to inconsistent findings. We also wanted to identify such instructional components that would best predict positive outcomes in intervention studies for preschool or kindergarten students. In addition, we examined the magnitude of the relationship between effect sizes and sample and methodological characteristics of selected primary studies.

The final goal of the present study was to evaluate the quality of research studies used and provide possible recommendations and future directions in this field.

CHAPTER 4

METHODOLOGY

Literature Search

I searched five major databases in the field of education and psychology –
PsychLit, ERIC, Academic Search Full Text Elite (EBSCO), Papers First, and
Dissertation Abstracts from 1977 to 2003. The following words and their
combinations were used: 3-6 year-olds, addition, arithmetic, calculators and
computers, computation, concepts (conservation) of number, counting, digit, division,
early childhood mathematics, enumeration, estimation, grouping, kids, kindergarten,
learning, mathematics, mathematical ability, math achievement, mathematics
concepts, measurement, meta-analysis, metrics, money, multiplication, number,
numerical reasoning, preschool education, probability, problem solving, quantity,
reasoning, sorting, standards, subtraction, teacher-child interaction, time, whole
number, writing. This search yielded approximately 2000 items that included articles,
book chapters, dissertations, and technical reports. All the abstracts were examined
and those materials that clearly did not meet the 3 main inclusion criteria outlined
below were excluded (e.g., if the sample in the study did not meet the age criterion). I
also searched the references of relevant books, chapters, and other articles on this
subject for additional titles.

Computer searches do not include unpublished studies. To locate such studies, I contacted the editors of the following journals: *Journal of Educational Psychology, the Arithmetic Teacher, Early Childhood Education, Cognitive Psychology, Journal for Research in Mathematics Education, Journal of Educational Research*, and asked them to identify the names of the researchers currently working in the field of early mathematics. We originally contacted all 60 of the nominated researchers who conducted research in the area of early mathematics and asked them to identify any training studies in mathematics in 3-6 year-old children. Additionally, we asked for copies of unpublished and/or current intervention studies and or names of other people who were working in the field of early mathematics. Sixty-eight percent responded to our e-mail messages. As a follow up, 12 people nominated by researchers were contacted. A year later we sent electronic mail to the same 72 researchers and, following up on their suggestions, we contacted additional 25 people.

Finally, we conducted a manual search of leading journals in mathematics education for the year 1970 through present. These journals were: *Arithmetic Teacher, Journal of Educational Psychology, Cognitive Psychology, Cognition and Instruction, American Educational Research Journal, Journal of Educational Research,* and *Child Development.* As a result of all these searches, we examined a total of 2371 reports, dissertations, articles, or chapters (see Table 2).

TABLE 2

SEARCH RESULTS

References/Abstracts Searched	Articles Examined According to 3 Criteria	Articles Examined According to 5 Additional Criteria	Studies Procured
2371	234	147	29

Selection Criteria

The pool of 2371 studies was narrowed down by examining the abstracts and

finding training studies that would fit the age range and the domain of interest.

Specifically, we looked at whether the studies met 3 main criteria: 1) the children in

the study were within the 3 to 6 age range at the onset; 2) the treatment group

received instruction/intervention that was more than their typical classroom

experience, and 3) children's mathematical knowledge or skills were assessed (see

Appendix A). This procedure yielded 234 studies that could be potentially included in

our meta-analysis. We then evaluated each of these 234 studies. We did not restrict

ourselves to any particular type of skill or any particular type of interventions. We

excluded 8.5% (n=20) of the 234 studies due to children in the studies exceeding the

age range restrictions (all those studies had combined samples of older children as

well as 3 to 6 year-olds), 17.5% (n=41) due to the absence of an intervention that met

our criteria, and 11.1% (n= 26) of the total number of studies because the main focus

was strategy (not mathematics skills per se) or teacher's behaviors. The remaining

147 studies were further coded according to additional 5 criteria: 1) the study had to

be in English; 2) the study had to be conducted after 1977; 3) the study had to report sufficient quantitative information to conduct a meta-analysis; 4) the study could not be a case, single-subject, or correlational study; and 5) the study had to have included a control group (See Appendix A). We excluded 6.8% (n=10) of the 147 studies because they were not in English and 6.1% (n=9) because they were conducted before 1977. We excluded 19% (n=28) of studies because they were case, single-subject, or correlational studies. We also excluded all studies that did not have a control group (n=55, 37.4%), did not adjust pretest differences (if they existed) (n=37, 25.2%), or did not have sufficient quantitative information (n=45, 30.6%). Out of the 45 studies that lacked quantitative information, there were 8 studies that did report some but not all necessary information to calculate the effect sizes (these eight studies satisfied all six postulated criteria). We contacted eight researchers and received additional information from two authors.

As a result, 31 studies were coded. In the process of effect size calculations and analyses, we also excluded 2 studies that did not use random assignment. A list of these 29 studies (19 published and 10 unpublished studies) that yielded 32 effect sizes (three studies had two separate experiments in them) is available in Appendix C.

Coding

Studies were screened for inclusion (2 coders) and coded (2 different coders) by research assistants to avoid the likelihood of experimenter expectancies. All of the coders had taken statistics and methodology courses in psychology. To create a coding scheme, we: 1) reviewed 10 previous meta-analytic intervention studies

(Baker et al., 2002; Bus & van Ijzendoorn, 1999; Elbaum et al., 2000; Fuchs et al., 2000; Gersten & Baker, 2000; Gersten et al., 2000; Marquis et al., 2000; Okolo et al., 2000; Swanson, 2000; Swanson & Lussier, 2001), 2) created a list of essential characteristics of empirical training studies based on accepted methodological and statistical research practices, and 3) reviewed books on meta-analysis. The creation of the intervention variables is described in a separate section below. This process yielded a total of 187 variables to code.

Three training sessions two hours each were conducted to help the two coders identify necessary information and code according to the developed criteria. The coders were then asked to code an intervention study in math and meet again to discuss any questions or difficulties. Coders identified different parameters of each criterion during this follow up as well as made revisions to how criteria were worded. Eight variables were deleted, five new variables were added, and 22 variables were modified. We randomly selected four papers on math intervention (not considered in this meta-analysis) and coded to see if (a) the coding criteria were well defined; (b) no essential elements had been accidentally omitted from the coding criteria; and (c) a set of intervention codes could be developed (described in details below). After the coding sheet was adjusted again (a total of 9 variables were added, nine variables modified, two variables deleted), each of the two coders read the articles independently and assessed them according to established criteria. Appendix B lists the 191 variables that were used in this meta-analysis.

The second coder coded 10 studies. Interrater reliability ranged from r=. 80 to r=1.00. After the coding, the protocols were compared. When some of the studies did not fit the established criteria well, the coders discussed the discrepancies in order to find the best possible solution. The agreed-upon protocol was entered directly into SPSS. Means and standard deviations were entered into Comprehensive Meta-Analysis software for the effect size computations. We describe our logic in calculating effect sizes in section "Effect Size Calculations".

All coded variables could be subdivided into four groups: study characteristics, sample description, intervention, and measurement information. The first group of study characteristics included, for example, such variables as author's name, university affiliation of the study, source type, year of publication, and country. The second group was sample description including the following variables as location of the sample, setting, sample size, information on age, gender, SES, and ethnicity. The third group was intervention and it included the following information: the number of intervention sessions, amount of time per week spent in training, frequency of sessions, language, program age, components of math instruction, instructed domains, types of skills tested and practiced during intervention, specific intervention techniques (described separately in the section called Intervention Variables), who administered intervention. The fourth group was measurement information and it included such variables as information on types of measures used, reliability, convergent and discriminant validity, test administrators, and number of items in each measure. A full coding sheet is included in Appendix B.

The development of the classification scheme for the intervention variables is described in the next section below.

We used two distinct approaches in developing a classification scheme for the intervention variables: induction and deduction. The first was an inductive approach in which we used intervention classifications used by previous literature reviews and researchers in the field of education. Four such classifications were obtained as specified in Table 1: by Dixon et. al (1998), by Swanson and Hoskyn (1998), by Maccini and Hughes (1997), and by Lee (Lee, 2000) (see details in Chapter 2). A combination of the above mentioned classifications was adopted. We borrowed from a classification scheme suggested by Dixon and colleagues. Instead of seven instructional characteristics (i.e., seven variables) suggested by these authors, we created one variable with seven possible response options. Therefore, each study received one characteristic depending on its research question. Twenty original variables were implemented as proposed by Swanson and Hoskyn (Swanson & Hoskyn, 1998). In addition, we added 5 variables from Lee's (2000) classification scheme: sequencing examples, adjusting difficulty levels, using peer tutoring, using motivation-based intervention, using mastery criterion, and using specific feedback.

The second was a deductive approach in which we reviewed 4 of the 31 selected training studies and made a list of instructional components used in interventions. In addition, we examined 128 intervention characteristics proposed in Dixon et. al. (Dixon et al., 1998). Based on interventions used in preschool and kindergarten and Dixon's list of instructional characteristics, we added the following list of instructional components: external rewards (individual external rewards, group external rewards, positive teacher attitudes, student team learning), feedback with corrections (yes or no), fluency building drills (yes or no), homework (yes or no), parental involvement (yes or no), children's literature (yes or no), games (yes or no), small group games to teach number concepts (yes or no), and pull-out settings (yes or no). The principle investigator together with the other PhD level researcher discussed each of the categories and how they corresponded to the intervention information in the selected studies. As a result, 17 variables were modified (see Appendix B). For example, we combined 8 variables suggested by Swanson and Hoskyn (Swanson & Hoskyn, 1998) into 4 due to difficulty in coding and close resemblance in content: sequencing and segmentation, preparation response and instruction, and teacher modeling and modeling, small group instruction and group instruction. If a study used multiple methods of teaching or fit into several criteria, multiple codes were assigned. We used 25 intervention variables (see Table 6, Chapter 5) that represented main instructional techniques in effect size analyses.

Classification of Instructional Approaches

Two coders read the description of the training for each study and rated each intervention as a whole. By definition, a direct instructional approach had at least 50% of the intervention following specific learning objectives usually with a protocol of activities, learning goals, criteria for mastery, and specific procedures to follow. A teacher usually modeled a given task, provided feedback and allowed opportunity for practice. Guided instruction often allowed for flexibility in the materials and procedures with the child taking the lead in the process of learning. An adult acted as a coach providing some help with no explicit modeling and constant repetitions. The agreement between coders was .87 (CI from .73 to .93). Those studies that the coders disagreed on were discussed until the agreement in assignment was reached.

Coding of the Dependent Variable

We coded each dependent measure either as skill specific or standardized. Skill-specific measures were closely related to the intervention manipulations. Some of the measures, for example, included tests on enumeration, test of comparison skills, geometrical reasoning. When there was more than one skill specific test in a given study that measured mathematics performance, we calculated effect sizes for each test and then reported mean effect size in the meta-analysis.

Standardized mathematics achievement measures usually had a number of mathematical skills and topics assessed including word problems, reasoning, and more difficult mathematical concepts such as fractions, division, etc. TEMA-2, SESAT, O-LSAT are examples of standardized tests that were used. These measures were not intervention specific and they assessed improvements in general mathematical competences and performances.

Quality Control

We created a quality index for each study based on the quality criteria used in different meta-analyses (Swanson, 2000). Each study received one point if it satisfied any of the following:

1) reported reliability;

2) reported validity information;

3) used random assignment;

4) when instruction was more than 10 sessions;

5) when measures of treatment integrity were used;

6) when outcome measures were standardized;

7) comparability in the control and experimental groups in curriculum was established;

8) comparability in the control and experimental groups in duration was established;

9) when counterbalancing was used.

The scores were summed across for a total index of quality. Those studies below the median fell into a category of " Low Quality" studies. Those studies above the median or higher fell into a category of "High Quality studies".

Effect Size Calculations

An *effect size* is a "degree to which a phenomenon is present in the population or the degree to which the null hypothesis is false" (Cohen, 1988, pp.9-10). We chose Cohen's d as the primary index of effect size of interest. It is usually calculated as a difference between the mean of the intervention and the mean of the control group divided by the pooled standard deviation of the two groups (Cook et al., 1992). For complex ANOVA designs we followed procedures outlined in Cortina and Nouri (Cortina & Nouri, 2000).

Effect sizes for between-groups comparison were based on the comparison of two posttest means from different groups divided by the pooled standard deviation. The pooled standard deviation was calculated as the average of the experimental posttest standard deviation and control posttest standard deviation. If homogeneity of variance could not be assumed, standard deviation of the control group was used instead of a pooled standard deviation (Cortina & Nouri, 2000; Glass et al., 1981). If means and standard deviations were not available, we estimated effect sizes based on ts, Fs, or ps (Cortina & Nouri, 2000). If a study did not report the ns for experimental and control conditions (and if ns could not be estimated based on statistics provided in the text), we divided the total N by two.

A positive effect size in a between-group comparison reflected improvement of the experimental group over the control group. In the event that the study yielded both experimental versus control group and experimental versus experimental comparisons, the first type of effect size was selected for inclusion in the meta-analysis. If a study had a dummy-treatment and a no-intervention control groups (e.g., Sophian, 2003), a control group with dummy treatment was preferred. We also calculated effect sizes for no-treatment controls and later conducted an analysis were the magnitude of the effect sizes was analyzed a function of the type of the control group.

If multiple effect sizes were possible, we used the following procedure. First, if a study had k number of treatment groups, an effect size comparing an average of the two treatment group means to the control was calculated resulting in one effect size per study. Second, if different dependent measures were used in the study (e.g., standardized test and a specific math skill test), effect sizes for each measure were obtained (also see below for details on the dependent variable coding). When separate publications of the same study and sample were identified, we treated them as the

49

same study, not as two separate studies (Pasnak, Madden et al., 1996; Pasnak et al., 1991).

Homogeneity

We checked homogeneity of study results to see whether the amount of variance in the observed set of effect sizes was significantly different from the amount of variance expected by sampling error alone. If Q was significant, the variability of effect sizes was greater than that which could be attributed to sampling error and the effect sizes were representative of different populations. We then conducted further analyses to discover if the variability could be explained by a particular coded variable.

CHAPTER 5

RESULTS

Appendix D provides an overview of each study, including the authors, year of publication, sample size, mean age, duration and description of the intervention, skills practiced during intervention, major findings for each study, and calculated effect sizes. We present the findings in three major sections. First, descriptive information on study, sample, and intervention characteristics is provided. Second, we address the primary research question on the impact of the type of training on children's development of mathematics skills. Finally, the effects of different variables on the magnitude of the effect sizes were explored.

Descriptive Information

Outliers. We examined scatterplots and boxplots to identify potential outliers (Lipsey & Wilson, 2000) as extreme effect sizes might have disproportional influence on measures of dispersion and central tendency. Graph 1 presents a boxplot of the distribution of the 32 calculated effect sizes.

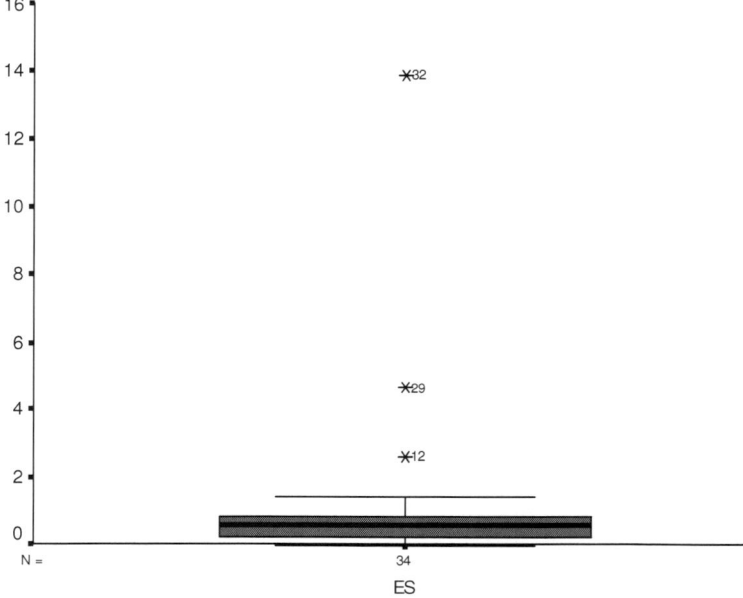

Figure 1. A Boxplot of Calculated Effect Sizes

The median for each dataset is indicated by the black center line, and the first and third quartiles are the edges of the box. We closely examined the entry level descritptive information in those studies and did not find any mistakes in effect size calculations. Three points were outside of the boxplots. Three studies were identified as outliers (studies 4, 17, 28 in Appendix C). As a result, these studies with effect sizes of d=4.77, 14.26 and 2.68 were excluded. The rest of the analyses were conducted excluding outliers from the set of analyzed studies. The main effect size calculations with the outlier in the model are presented in Appendix D.

Study and Sample Characteristics. After removing the outliers, we analyzed

29 studies that came from 26 sources. Appendix D reports the authors, year of

publication, sample size, mean age, duration and description of the intervention, skills

practiced during intervention, major findings for each study, and calculated effect

sizes. The average year of publication for studies was 1993 (SD=7.91 years) with the

range of publication dates from 1978 to 2003. 62.1% of studies (n=18) were

published articles.

A total of 1845 children participated as subjects in the 29 studies (n=999 in

the experimental group; n=846 in the control group). 75.8% of the studies drew

participants from US samples. The sample sizes for each study are presented in

Appendix D. Sample size ranged from 18 to 160 participants per study with a mean

sample size of 77 subjects per study (SD=38.64). Of the 29 coded studies, 12 studies

(41.4%) involved kindergarten students and 17 studies involved (58.6%)

preschoolers. As a result, subjects in the studies ranged in age from 48.25 to 74

months with a mean of 57.68 months and a standard deviation of 7.65 months. Less

than half of the studies reported information on ethnicity, SES, and gender. Twenty

studies (69%) reported participants' gender. On average, 53% of the participants were

male. Of the 20 studies that reported SES, 30% drew participants from middle-class,

45% from lower SES, and 25% had mixed samples. Two studies exclusively studied

middle-class children and 8 studies drew participants exclusively from low SES.

Locale of the study was urban (40.7%), suburban (14.8%), mixed (3.7%) or unknown

(40.7%). Response rate ranged from 61.40% to 100%. 100% of the studies used

random assignment.

Instructional Characteristics. A typical intervention study included on average 1.16 hours of instruction per week ranging from 1 to 5 times a week (*M*=2.88 sessions, *SD*=1.48) over 1 to 36 weeks. On average, training lasted for 35 sessions (*SD*=47.61) ranging from 1 to 180 sessions. 82.8% of the studies provided instruction in English. 31% of the studies used an established intervention programs (e.g., Bright Start). According to the type of instructional approach used, we classified 9 studies as guided (24.1%), 9 (31%) as direct, and 13 (44.8%) as mixed. 33.3% of the total number of coded studies provided instruction in a classroom setting. Children learned about mathematics through a variety of means: print, games, technology, and activities. On average, about 29% of all coded studies used books in their interventions, 53.6% used games, 10.7% used worksheets, 10.3% used technology, and 75% used activities.

Table 3 provides percentages of the total number of studies that assessed or taught different math skills.

TABLE 3

PERCENTAGES OF THE TOTAL NUMBER OF STUDIES THAT ASSESSED OR

TRAINED PARTICIPANTS ON DIFFERENT MATHEMATICS SKILLS

Skills	Assessed	Trained
Number id	17.2	17.2
Number sense	31.0	20.7
Counting	51.7	48.3
Estimation	3.4	10.3
Computation	34.5	41.4
Classification	24.1	27.6
Ordinality	3.4	17.2
Cardinality	10.3	13.8
Writing numerals	10.3	3.4
Number-object correspondence	37.8	41.4
Comparison	44.8	37.9
Geometry	20.7	24.1
Measurement	17.2	13.8
Fractions	3.4	3.4
Conservation	34.5	44.8

24% of all the studies also included a delayed posttest measure to check for maintenance of acquired skills. 17.2% of all the studies investigated transfer performance.

There were a number of problems in the coded studies (e.g., with instructors, with intervention descriptions, and with measures used). Only 44.8% of the total number of coded studies explicitly described how they had trained their instructors. In

18 studies (62.1%) it was not clear whether instructors were assigned to conditions randomly or not. In 4 studies (13.8% of the total number of studies) instructors alternated between a treatment and control groups, while in 7 studies (24.1%) the same instructor was assigned to a specific condition. In 18 studies (62.1%) information on instructor variability was not reported.

Full description of the interventions was available in 58.6% of the total number of studies, 37.9% of the total number of coded studies provided general description, and 3.4% of the total number of studies used descriptive labels for their intervention procedures. On average across 17 coded studies, anywhere from 0 to 66% of participants withdrew from the experimental group after the beginning of the project, while attrition in the control group ranged from 0 to 47.5%. Twelve (41.4%) studies did not report information on attrition.

In 21 studies (72.4%) the information on test administrators was not reported. Less than half of the studies provided information on reliability and validity of measures (for details see section 5.3 High and Low Quality Studies). Only 4 studies (13.8%) controlled for Type I error. Due to significant quality differences that existed between the studies, all studies were divided into high and low quality studies. We investigated the magnitude of the effect sizes as a function of study quality in section "Effects of Different Variables on the Magnitude of the Effect Sizes."

Between-Group Comparisons

After removing the outliers, we had 26 independent sources that included 29 studies (3 sources had two separate experiments in each one of them). The estimated effect sizes, in terms of an unbiased *d* score, for between-group comparisons for each of the studies are presented in Appendix D. The following information is provided in this appendix: the authors and year of publication, sample size, mean age, duration and description of the intervention, skills practiced during intervention, major findings for each study, and calculated effect sizes. We conducted a meta-analysis of these individual effect sizes using statistical software "Comprehensive Meta-Analysis".

The mean adjusted and unadjusted effect sizes for all studies for both fixed and random models are presented in Table 4. We used Hedge's g as a correction method for upward bias in effect sizes (Lipsey & Wilson, 2000).

TABLE 4

MEANS OF EFFECT SIZES

Index	Standardized Difference (d) and 95% Confidence Interval			
	Point Estimate	SE	Lower Limit	Upper Limit
Cohen's d (unadjusted)				
Fixed	.488	.047	.395	.581
Random	.497	.066	.367	.627
Hedge's g (adjusted)				
Fixed	.467	.048	.372	.562
Random	.478	.064	.353	.603

NOTE: Number of outcomes=29. Number of individual cases=1845.

57

Effect sizes ranged from -.047 to 1.389. The mean effect size across all studies was .467, $p<.01$ (*CI* from .372 to .562) for a fixed effect and .478, $p<.01$ (*CI* from .353 to .603) for a random effect. Thus, the between-group comparisons showed the benefits of training over a control condition.

When a weighted regression framework was used to see whether teaching children a specific skill (e.g., counting) would moderate effect size estimates, we found that only instruction on measurement was a significant negative predictor ($B= -.36$, $p<.05$). Teaching other mathematics skills did not have an effect on the magnitude of the effect sizes.

The test for homogeneity for the overall model yielded a significant Q statistic of 44.81, $p < .05$. Thus, although the effect size illustrated a significant difference between experimental and control conditions, the effect sizes making up the meta-analysis were widely different. Additional analyses were needed to help explore these differences and to try to uncover the variables responsible for them.

Effects of Different Variables on the Magnitude of the Effect Sizes

In the absence of homogeneity, the coded categorical variables were used to account for this variability. Categorical model testing, which is analogous to univariate analysis of variance, divides the study pool into subgroups based on selected study characteristics. Each model calculates the statistic Q_B, which had a chi-square distribution of $p -1$ degrees of freedom, where p is the number of classes. A significant Q_B indicates that effect sizes differ between subgroups. We first explored whether the type of instructional approach (guided, direct or mixed) was related to the

58

magnitude of the effect sizes. Second, we analyzed how different instructional techniques contributed to the magnitude of the difference between the experimental and control groups. Third, we analyzed the magnitude of the effect sizes as a function of the type of measure used. Finally, we report the results of the analyses of such variables as the length of an intervention and the quality of analyzed studies. Note that additional variables of interest (e.g., the type of the analyses used, the type of the control group) were explored but no differences in the magnitude of effect sizes were obtained.

Types of Instructional Approach

Based on the type of instructional approach used, we classified 7 studies as guided (25.9%), 7 (25.9%) as direct, and 13 (48.1%) as mixed (Two studies with both approaches were excluded from this analysis). Effect sizes for each type of approach are presented in Table 5.

TABLE 5

MEANS OF EFFECT SIZES FOR THREE INSTRUCTIONAL APPROACHES

Index	Standardized Difference (d) and 95% Confidence Interval				
	N Point Estimate	SE	Lower Limit	Upper Limit
Hedges's g				
Guided	7 .428	.084	.263	.594
Direct	7 .154	.101	-.045	.352
Mixed	13 .630	.076	.480	.779
Combined	27 .480	.049	.351	.545

According to the 95% confidence intervals, all approaches produced non-zero effect sizes. A significant $Q_B = 14.24$, $p<.01$ indicated that effect sizes differed between subgroups (i.e., guided, direct, and mixed) by more than sampling error. The residual variability, $Q_W = 20.67$, $p>.05$ could be considered homogenous. Further analyses indicated that there was a significant differences in effect sizes between direct and guided approaches ($Q_B = 14.24$, $p<.01$) and direct and mixed approaches ($Q_B = 14.16$, $p<.01$) but not between effect sizes for mixed and guided approaches ($Q_B = 3.14$, $p<.10$). Table 5 presents weighted mean effect sizes for a fixed model for each of the instructional approaches as well as for the combined model. Two findings are notable from the analyses. First, a combination of both guided and direct approaches as opposed to either a direct or a guided approach by itself yielded the highest effect size and could therefore, be considered the most beneficial for teaching young children mathematics. Second, an effect size of .428 for a guided approach as opposed to .154 for a direct approach suggests that children at an early age benefit from child-centered approach a lot more than from a teacher-centered approach ($Q_B = 4.36$, $p<.05$).

Specific Instructional Techniques

Table 6 shows weighted effect sizes for fixed effect models for each of the instructional methods. The first column indicates different instructional methods (ordered by the magnitude of the mean weighted effect size) used. The second column indicates a number of studies that used a specific instructional method. A weighted mean effect size for a fixed model is reported in the third column. The fourth column reports standard error for this mean weighted effect size. The fifth and sixth columns report the 95% confidence intervals for each mean weighted effect size.

TABLE 6

SUMMARY OF AGGREGATE EFFECT SIZES FOR DIFFERENT

INTERVENTION CATEGORIES

Category	No. of Studies	Participants' N	WES	SE	95% Confidence Interval for WES	
					Lower	Upper
Peer tutoring	3	109	.708	.198	.315	1.102
Controlling task difficulty	14	781	.663	.074	.518	.810
Elaboration	11	799	.605	.075	.458	.753
Sequencing and/or segmentation	14	898	.593	.070	.457	.730
Fluency building	3	119	.578	.188	.205	.951
Reinforcement	12	777	.576	.074	.430	.721
Small group games	15	1086	.570	.063	.446	.693
Strategy	10	626	.549	.083	.387	.711
Novelty	8	678	.544	.080	.388	.701
Modeling and/or teacher modeling	14	851	.532	.073	.389	.675
Mastery criterion	7	364	.528	.110	.316	.740
Group instruction	21	1509	.520	.053	.415	.624
Dialogue	17	1128	.516	.063	.393	.640
Motivation-based intervention	5	384	.508	.105	.302	.714
Preparation responses and/or instruction	14	862	.497	.072	.355	.639
Pull-out setting	17	972	.492	.068	.359	.625
External rewards	10	632	.489	.082	.327	.651
Drill-repetition/practice	10	544	.450	.091	.272	.628
Nonteacher instruction	9	685	.420	.078	.267	.574
Feedback	14	827	.411	.072	.271	.552
Supplement	7	529	.361	.089	.187	.535
Homework	5	338	.354	.111	.136	.573
Parental involvement	6	443	.322	.100	.132	.513

TABLE 6 (continued)

Category	No. of Studies	Participants' N	WES	SE	95% Confidence Interval for WES	
					Lower	Upper
Technology	3	223	.304	.137	.034	.574
One-to-one instruction	17	955	.270	.067	.138	.403

According to the third column, the studies that favored interventions including peer tutoring, controlling task difficulty, additional explanations provided about taught concepts, and sequencing activities produced the largest effect sizes. One-to-one instruction, technology, including parents, and homework were less successfully used instructional techniques. The magnitude of the effect did not coincide with the frequency of examining a specific technique. For example, pull out setting was examined in 17 studies, yet it produced a relatively smaller effect sizes than other techniques.

We also identified those categories that were significant predictors of effect size estimates. Using a WLS regression framework and an SPSS macro developed by Wilson, each of these instructional categories was entered into the equation as an independent variable. The results indicated that variables such as one to one instruction (it explained 33% of variance in the effect size estimates), controlling task difficulty (35% of variance explained), group instruction (19%), elaboration (14%), and sequencing (14%) were significant predictors.

Standardized and Skill-Specific Measures

Standardized tests of mathematics tend to have better established reliability and validity than skill-specific researcher-created tests. 29% of the total number of studies used standardized measures, 62.1% - skill-specific measures, and 6.9% - used both types. We excluded those studies that had both types of measures and ran separate analyses for standardized (9 effect sizes) and skill-specific measures (18 effect sizes) to analyze the possible effects of test type.

The magnitude of mean weighted effect sizes was lower for the skill-specific measures than for standardized measures of math achievement (see Tables 7 and 8). The significant Q_B statistics of 7.05, $p<.05$ indicated that effect sizes were in fact different for two types of measures.

TABLE 7

MEANS OF EFFECT SIZES FOR STANDARDIZED MEASURES

Index	Standardized Difference (d) and 95% Confidence Interval			
	Point Estimate	SE	Lower Limit	Upper Limit
Cohen's d (unadjusted)				
Fixed	.656	.071	.502	.811
Random	.669	.101	.471	.866
Hedge's g (adjusted)				
Fixed	.635	.073	.476	.795
Random	.644	.094	.459	.828

NOTE: Number of outcomes=9. Total number of subjects=652

64

TABLE 8

MEANS OF EFFECT SIZES FOR SKILL-SPECIFIC MEASURES

Index	Standardized Difference (d) and 95% Confidence Interval			
	Point Estimate	SE	Lower Limit	Upper Limit
Cohen's d (unadjusted)				
Fixed	.381	.064	.256	.505
Random	.395	.088	.222	.569
Hedge's g (adjusted)				
Fixed	.360	.065	.233	.487
Random	.377	.074	.218	.542

NOTE: Number of outcomes=18. Total number of subjects=1052.

Length of Intervention

We conducted a weighted regression analysis to see if the magnitude of the effect sizes could be predicted from the length of intervention (measured as the number of instructional sessions). It was not true for our data that studies with longer interventions produced higher effect sizes (Q_B= 3.47, p>.05). The length of the intervention was not a significant predictor of the estimates of effect size (B= .33, p>.05).

High and Low Quality Studies

All studies were classified as either of low or high quality. The nine variables that were used to determine whether a study was of a low or a high quality were: 1) reported reliability; 2)reported validity information; 3) random assignment; 4) when instruction was more than 10 sessions; 5) when measures of treatment integrity were used; 6) when outcome measures were standardized; 7) when comparability in the control and experimental groups in curriculum was established; 8) when comparability in the control and experimental groups in duration was established; and 9) when counterbalancing was used. Each study that was above the median (more than 4 points) was considered of high quality (n=10 studies).

Below we present descriptive information on the main criteria used in this categorization. 51.7% of the total number of studies taught children for more than 10 sessions. 31% of all the studies used counterbalancing. 44.8% of the studies employed manipulation checks. Experimental and control treatments were comparable: (a) in duration in 65.5% (n=19) of all the studies; (b) in instructors in 41.4% (n=12) of the studies; and (c) in curriculum in 69% (n=20) of the studies. 12 studies (41.4%) provided information on reliability of used measures. 6 studies (20.7%) reported information on convergent validity. Study quality was not related to whether a study was a published one or not ($t(27)$= -.23, p>.05).

Table 9 shows the weighted effect sizes for low and high quality studies. In general, high quality studies produced an effect size twice the magnitude of the effect size produced by low quality studies. The difference between mean weighted effect sizes was statistically significant ($t(28)$=-2.61, p<.05).

66

TABLE 9

MEANS OF EFFECT SIZES FOR HIGH AND LOW QUALITY STUDIES

Index	Standardized Difference (d) and 95% Confidence Interval					
	N	Point Estimate	SE	Lower Limit	Upper Limit	Q
Hedges's g						
High	10	.581	.063	.457	.705	20.45
Low	16	.301	.076	.152	.450	16.28
Combined	26	.467	.048	.372	.562	44.81*

NOTE: * $p < .05$

Additional Variables

We ran weighted regression analyses to check whether such variables as the year of publication, sample size, whether a study was published or not, number of weeks of treatment, age, type of the control group, type of design, type of population employed, percentage of attrition significantly contribute to the prediction of effect size estimates. When sample size was analyzed, an unweighted effect size was used as weighting is based on the reciprocal of sample size. The results indicated that none of these variables were significant predictors.

CHAPTER 6

DISCUSSION

Is Mathematics Instruction Beneficial?

Our findings provide evidence of effectiveness of early mathematics instruction. The current meta-analysis shows that on average early intervention in mathematics positively influences the magnitude of the effect sizes (i.e., a standardized difference between an experimental and a control group). We found an overall unadjusted mean effect size to be .49. The magnitude of the effect size was not substantially reduced when Hedge's correction was applied (g=.467). On average across all types of interventions and students, early mathematics instruction was effective.

This obtained effect size corresponds to Cohen's medium effect size and to a correlation effect size of about .24. Often an effect size is understood in terms of the Binomial Effect Size Display when the success rate of the control and treatment groups are compared. The standardized difference of .467 can be described as a success rate of 68% in the treatment group compared to a success rate of 50% in the control group. That is, 68% of the treated participants are above the control group median, while only 50% of participants in the control group are above that level.

Our weighted mean effect size is comparable to the overall mean effect size after correction for sample size in other meta-analyses. Xin and Jitendra (1999) reported an overall mean effect size of .89 (which is a different measure of effect size as it was operationalized as a standardized pre-post change in performance). Swanson reported an overall effect size of .79 for students with learning disabilities (Swanson, 2000) but her meta-analysis was not constrained to math interventions. When she looked at 28 studies in mathematics, the weighted effect size dropped to .40.

This meta-analysis of 29 studies also provided some information on what techniques and approaches contribute to improvement in children's math performance.

Guided or Direct Instruction?

One consistent finding is that a combination of guided and direct approaches to teaching mathematics seems to enhance students' performance. The interventions using this mixed approach showed the highest effect sizes. While the debate on whether to use guided child-centered approach or direct teacher-centered approach still continues, it is clear that each of those approaches in isolation produced smaller effect sizes than when mixed.

The other interesting finding is that young children seem to learn better from a guided approach than from a direct teacher-centered approach as the effect sizes produced from a guided approach were consistently larger than effect sizes produced from a direct approach. This latter finding contradicts findings from other studies where direct instruction was more efficient (Kroesbergen & Van Luit, 2003). A possible explanation could be that in our analysis we did not control for the type of population under investigation while other studies found direct instruction to be successful if students were low achieving or had special needs.

Specific Instructional Techniques

We followed Swanson and Hoskyn's approach (Swanson & Hoskyn, 1998) and coded for different details of components of effective instruction as not all forms of intervention work equally well. Results clearly support the notion that treatment effects, especially, those studies that employed peers as tutors, controlling task difficulty, additional explanations provided about taught concepts, and sequencing activities contributed independent systematic variance to the magnitude of the effect sizes. This finding is in line with Vygotsky's approach to teaching children mathematics where tasks are broken down and adjusted to the learner's level of development. Effectiveness of peer tutoring contradicts Kroesbergen and Van Luit's meta-analysis findings who reported lower effect size for peer assisted performance in mathematics (Kroesbergen & Van Luit, 2003). Their meta-analysis was on elementary children with special needs. In Baker and colleague's meta-analysis on low-achieving students, peer tutoring led to positive effects on student achievement with an effect size of .62 (Baker et al., 2002).

Another conclusion we can draw is that providing one-to-one instruction by itself, using technology, parental involvement or homework were not as effective. The finding of limited effectiveness of technology corresponds to the findings from other studies (e.g., Kroesbergen & Van Luit, 2003) where using technology was less effective than a teacher. However, Xin and Jitendra (1999) reported computer-assisted instruction to be the most effective. A possible explanation for our finding is that on one hand, very few studies (n=3) employed technology. Additionally, most of Xin and Jitendra's studies used direct instruction which in their meta-analysis produced high effect sizes and as a result was confounded with the use of technology.

We also found that a few instructional techniques increase the predictive power of treatment effectiveness. Effect size estimates could be reliably predicted from the following instructional techniques (from 14 to 33% of variance in effect size estimates was accounted for depending on a specific predictor):

1. one to one instruction (activities related to independent practice, tutoring, instruction that is individually paced, and/or instruction that is individually tailored),

2. adjusting difficulty levels (changing difficulty levels of material or tasks to accommodate learner characteristics),

3. group instruction (instruction in a small group, and/or verbal interaction occurring in a small group with students and/or teacher),

4. elaboration (additional information or explanation provided about concepts, procedures or steps, and/or redundant text or repetition within text), and

5. sequencing breaking down the task, fading of prompts or cues, matching the difficulty level of the task to the student, sequencing short activities, and/or using step-by-step prompts; breaking down the targeted skill into smaller units, breaking into component parts, segmenting and/or synthesizing components parts).

Problems with Methodology

There were a number of characteristics that were not reported in our coded studies (e.g., with intervention descriptions, with instructors, and measures used). About 40% of studies provided information on instructors, their variability and training. Only 31% of all the studies used counterbalancing. Less than half (44.8%) of the studies employed manipulation checks. Very often the studies were not comparable in instructors or curriculum. Only 12 studies (41.4%) provided information on reliability of used measures. 6 studies (20.7%) reported information on convergent validity. None of the studies reported discriminant validity information.

The finding that study quality did not depend on whether it was a published study or not points to the fact that even published articles have problems with such important methodological issues. As a result, efforts should be made to improve these aspects of research in educational psychology. One way to do so is to be more rigorous in the design and implementation research; another method is to clearly provide details of intervention, measures used, and other study related variables.

Som   e methodological characteristics contributed to treatment outcomes.

1. Studies that used standardized math measures yielded on average higher effect sizes than skill-specific measures of mathematics.

2. Studies of high quality yielded significantly larger effects than studies of lower quality.

Implications for Practice

A number of conclusions for teachers and parents could be drawn as a result of these findings. Teaching mathematics should be attempted in preschool and kindergarten years on a consistent basis. It is often the case that such early instruction is not consistent as it might be viewed as too difficult for children to comprehend. Our results indicate that on average across all techniques and types of students, early mathematics instruction is beneficial.

Another finding concerns the instructional approach educators can use in teaching. While debates about direct teacher-centered and guided child-centered instruction are heated, it is important to realize that a combination of these two approaches is feasible.

We have also identified a set of specific skills and instructional techniques that are often used in teaching young children about mathematics. While every classroom is unique, we were able to estimate what instructional approaches were more beneficial for younger children. One to one instruction, an individualized mode of learning, for example, might not be as effective with younger learners as previously has been thought. We encourage educators to use a variety of instructional techniques, including peer tutoring, adjusting levels of difficulty, segmenting, and elaboration in their work.

Finally, due to the fact that standardized math measures contributed to higher effect size estimates, we suggest teachers use assessment in their work employing more reliable standardizes measures of math performance when possible.

Limitations

The present study has a number of limitations. In our analyses we only identified the influence of single predictors on effect size estimates as opposed to multivariate models. We restricted ourselves to univariate relationships mainly due to limited prior meta-analytic research that models the magnitude of the relationship between the effect size estimates and different instructional characteristics in younger students controlling for other variables. In addition, while we investigated whether the study quality affected treatment outcomes, we did not control for the quality in most of our analyses. While some researchers argue that such correction for methodology should not be attempted, it seems that controlling for study imperfections is important.

Our analyses used a fixed model approach because we assumed that effect size heterogeneity is due to the effect of different moderator variables. While there is a continuing debate on how to choose from a fixed, random or mixed effects model, we chose a fixed effect model due to: 1) its increased power to detect moderator relationships (as opposed to a mixed model); 2) minimum difference in our meta-analysis between random and fixed models, and 3) software and meta-analysis macros are only written for either fixed or random effects. Additional research is needed to better understand the effects of these decisions on meta-analysis findings.

Future Directions

Mathematics achievement in the United States is known to be lower than in other countries (Geary, 1995; Stevenson et al., 1993). Exposing children to mathematics instruction early on seems to be a natural step to address this difficulty. While there have been a number of recent developments in the area of literacy that promoted children's literacy training at a very early age (e.g., phonological awareness concept has been put forward in the last decade), early training in mathematics has not received such attention. Head Start centers, for example, have recently (2002) switched to a new nationwide program called STEPs with major emphasis on literacy instruction. No mathematics curriculum has been adopted in that program or as a separate program. Clearly, this is a one major indication that the country is not fully aware of the importance of mathematics training in preschool years. Teaching mathematics early is a social and political issue that needs to be further promoted at the government and national levels. Researchers need to be more vocal about the benefits of early mathematics instruction and take a leading role in policy decision makings.

While intervention studies in mathematics with younger children do exist, research interventions might not be easily adopted in classrooms. Presently, primary intervention studies do not provide such important information as the cost of materials, specific descriptions of activities, etc. that will help educators adopt research-based findings. Reporting extensive details of intervention projects is critical and cannot be underestimated both for consumers as well as conducting quantitative literature synthesis.

Through the process of searching for appropriate intervention studies with young children, we had to exclude a number of studies for methodological reasons. While some aspects of the study are more difficult to control than others, random assignment of participants to conditions, use of counterbalancing and manipulation checks and the presence of a control group are important study characteristics to think about at the design stage of the intervention project. They contribute to the overall high internal validity and affect every aspect of the study findings.

Meta-analysis is a technique that has become quite popular in recent years. Despite the fact that more and more meta-analytic studies of intervention are conducted, there is a shortage of work on the theoretical basis of this analytical technique. Most of the current advances can only be found in very specialized journal articles. More textbooks and applied manuals are needed to better understand different options a researcher faces. For example, every meta-analyst needs to decide what assumptions in terms of effects model to adopt. Overall, researchers are slow in investigating what type of model assumptions are favorable for what types of effect models. New advances both in theory and software development are needed to address this important issue. In addition, future meta-analyses of intervention studies with young children might employ a methodology where such variables as the type of instructional approach or level of quality are controlled for.

Conclusion

Our quantitative synthesis of the existing literature indicates that the effects of early exposure to mathematics instruction are in the medium range adopting Cohen's classification. The types of training that are beneficial for what types of children under what conditions is a question that has not been thoroughly investigated in younger children. We hope that learning more about young children's cognition (and as it pertains to learning mathematics) will eventually lead to reevaluation of the mathematics curriculum both at preschool and elementary levels with a goal to create such environments that will have a lasting effect on children's mathematics knowledge, skills, interest in math and eventually success in this area.

APPENDIX A:

STUDY SELECTION SHEET

DECISION ON INCLUSION:

ID_____

1. Are the children in the study within the 3 to 6 age range at the onset? YES
    NO

2. Did the treatment group receive instruction/intervention that was more than
    their typical classroom experience? YES NO

3. Were children's mathematical knowledge or skills assessed (no teacher's
    behavior, attitudes or strategy use) YES NO

ADDITIONAL:

1. Is the study in English? YES NO

2. Was it conducted after 1970? YES NO

3. Is it a case, single-subject , or correlational study? YES NO

4. Does a study have sufficient quantitative information to conduct a meta-
    analysis?

    LIST_____

5. Does a study have a control group? YES NO

APPENDIX B

CODING PROTOCOL

TABLE B.1

CODING PROTOCOL

CATEGORY	CODE OPTIONS	DEFINITION OF CODE
1. Name of the article	Title.	
2. Author's name	Name.	
3. University of…	University affiliation of study.	
4. University 2	Additional university affiliates.	
5. Affiliation	a. Psychiatrist, psychologist, counselor.	Professional affiliation of the first author.
	b. Academic.	
	c. Government agency.	
	d. Program agency.	
	e. Research firm.	
	f. Other.	
6. Funding	a. Agency/organization.	Source of research funding.
	b. Federal.	
	c. State/local government.	
	d. Funded, unknown source.	
	e. No funding indicated.	
	f. Internal university grant or award.	
	g. Federal and internal	
7. Journal	Name of journal.	
8. Source type	a. Journal.	Source of study.
	b. Conference paper.	
	c. Manual.	
	d. Dissertation.	
	e. Book/book chapter.	
	f. Technical report.	
	g. Government publication.	
	h. Unpublished.	
	i. Other.	
9. Year of publication	Year.	
10. Country	Country of study.	
11. ID number		Study ID number.

TABLE B.1 (continued)

CATEGORY	CODE OPTIONS		DEFINITION OF CODE
12. Locale	a.	Urban.	Locale of study.
	b.	Suburban.	
	c.	Rural.	
	d.	Mixed.	
13. Intervention setting	a.	Regular education classroom.	Setting of intervention.
	b.	Special education classroom (resource room, self contained room, etc).	
	c.	Individual or group tutoring.	
	d.	Computer lab only.	
	e.	Computer lab combined with classroom lessons.	
	f.	Laboratory or clinic.	
	g.	Special workshop or summer school.	
14. Practice number id	a.	No.	Did the intervention practice number id?
	b.	Yes.	
15. Test number id	a.	No.	Did the study test number id?
	b.	Yes.	
16. Practice number sense	a.	No.	Did the intervention practice number sense?
	b.	Yes.	
17. Test number sense	a.	No.	Did the study test number sense?
	b.	Yes.	
18. Practice counting	a.	No.	Did the intervention practice counting?
	b.	Yes.	
19. Test counting	a.	No.	Did the study test counting?
	b.	Yes.	
20. Practice estimation	a.	No.	Did the intervention practice estimation?
	b.	Yes.	
21. Test estimation	a.	No.	Did the study test estimation?
	b.	Yes.	
24. Practice computations	a.	No.	Did the intervention practice whole number computations?
	b.	Yes.	
25. Test computations	a.	No.	Did the study test whole number computations?
	b.	Yes.	
26. Practice classification	a.	No.	Did the intervention practice classification?
	b.	Yes.	

81

TABLE B.1 (continued)

CATEGORY		CODE OPTIONS	DEFINITION OF CODE
27. Test classification	a.	No.	Did the study test classification?
	b.	Yes.	
28. Practice ordinal numbering	a.	No.	Did the intervention practice ordinal numbering?
	b.	Yes.	
29. Test ordinal numbering	a.	No.	Did the study test ordinal numbering?
	b.	Yes.	
30. Practice cardinality	a.	No.	Did the intervention practice cardinality?
	b.	Yes.	
31. Test cardinality	a.	No.	Did the study test cardinality?
	b.	Yes.	
32. Practice writing numerals	a.	No.	Did the intervention practice writing numerals?
	b.	Yes.	
33. Test writing numerals	a.	No.	Did the study test writing numerals?
	b.	Yes.	
34. Practice number/object correspondence	a.	No.	Did the intervention practice number/object correspondence?
	b.	Yes.	
35. Test number/object correspondence	a.	No.	Did the study test number/object correspondence?
	b.	Yes.	
36. Practice comparison	a.	No.	Did the intervention practice comparison?
	b.	Yes.	
37. Test comparison	a.	No.	Did the study test comparison?
	b.	Yes.	
38. Practice geometry	a.	No.	Did the intervention practice geometry?
	b.	Yes.	
39. Test geometry	a.	No.	Did the study test geometry?
	b.	Yes.	
40. Practice measurement	a.	No.	Did the intervention practice measurement?
	b.	Yes.	
41. Test measurement	a.	No.	Did the study test measurement?
	b.	Yes.	
42. Practice fractions	a.	No.	Did the intervention practice fractions?
	b.	Yes.	
43. Test fractions	a.	No.	Did the study test fractions?
	b.	Yes.	

TABLE B.1 (continued)

CATEGORY	CODE OPTIONS		DEFINITION OF CODE
44. Practice conservation	a.	No.	Did the intervention practice conservation?
	b.	Yes.	
45.	a.		
46. Test conservation	a.	No.	Did the study test conservation?
	b.	Yes.	
47. Practice other skills			What other skills did the intervention practice?
48. Test other skills			What other skills did the study test?
49. Intervention sessions	Number of intervention sessions.		
50. Sessions' time	Hours/wk.		Amount of time of intervention per week.
51. Frequency of sessions	a.	Daily.	Frequency of intervention sessions per week.
	b.	2-4 hours per week.	
	c.	1-2 hours per week.	
	d.	Less than weekly.	
	e.	Once.	
52. Language	a.	English.	Primary language of instruction.
	b.	Spanish.	
	c.	Other.	
53. Program/intervention age	a.	New (<2yrs).	Established age of intervention program.
	b.	Established.	

TABLE B.1 (continued)

CATEGORY		CODE OPTIONS	DEFINITION OF CODE
54. Math domains	a.	Basic computations (+, -, *, /).	Math domains covered by teachers/interventionists during the intervention.
	b.	Problem solving.	
	c.	Reasoning.	
	d.	Computation application (computation with fraction, decimals, measurement skills).	
	e.	Understanding concepts (basic number, place value, operation on rational number, quantity or volume number, conservation or classification).	
	f.	Geometry.	
	g.	Conservation.	
	h.	Mixed.	
55. Components of math instruction	a.	Use of manipulative.	-Work among peers: cooperative, team-based, collaborative learning.
	b.	Work among peers.	-Design of instruction: discovery vs didactic instruction; training heuristics and strategies – thought process, thinking strategies, metacognitive strategy training; sequencing; mastery learning; other.
	c.	Design of instruction.	
	d.	Computers and technology.	-Computers and technology: calculators, logo, computers.
	e.	Grouping.	Idiosyncratic studies: CMB; children's literature; lattice algorithm; part-part-whole; homework; matching amount of instruction to learner preferences for amount of instruction; teaching complex fractions; teacher affect; using mastermind to teach logic; over program to teach transformational geometry; proof construction.
	f.	Reinforcement or motivation systems.	
	g.	Idiosyncratic studies.	
56. Sequencing and/or segmentation	a.	No.	Statements in the treatment description about breaking down the task, fading of prompts or cues, matching the difficulty level of the task to the student, sequencing short activities, and/or using step-by-step prompts; breaking down the targeted skill into smaller units, breaking into component parts, segmenting and/or synthesizing components parts.
	b.	Yes.	
57. Drill-repetition/practice	a.	No.	Practice reviews: statements in the treatment description related to mastery criteria, distributed review and practice, using redundant materials or text, repeated practice, sequenced reviews daily feedback, and/or weekly reviews.
	b.	Yes.	

84

TABLE B.1 (continued)

CATEGORY		CODE OPTIONS	DEFINITION OF CODE
58. Preparation responses and/or instruction	a. b.	No. Yes.	Anticipatory or preparation responses: statements in the treatment description related to asking the child to look over material prior to instruction, providing information to prepare student for discussion, and/or stating the learning objective for the lesson prior to instruction. Advanced organizers: statements in the treatment description about directing children to look over material prior to instruction, children directed to focus on particular information, providing prior information about task, and/or the teacher stating objectives of instruction prior to commencing.
59. Dialogue	a. b.	No. Yes.	Structured verbal teacher-student interaction: statements in the treatment description about elaborate or redundant explanations, systematic prompting students to ask questions, teacher and student talking back and forth dialogue, and/or teacher asks questions that are open-ended or directed.
60. Novelty	a. b.	No. Yes.	Statements in the treatment description about the use of developed diagrams or picture presentations, specialized films or videos, instruction via computers specification that new curriculum was implemented, and/or emphasis on teacher presenting new material from the previous lesson.
61. Modeling and/or teacher modeling	a. b.	No. Yes.	Strategy modeling + attribution training: statements in the treatment description about processing components or multisteps related to modeling from the teacher; simplified demonstrations modeled by the teacher to solve a problem or complete a task successfully; teacher modeling; teacher providing reminders to use certain strategies, steps, and/or procedures; think-aloud models; and/or the teacher presenting the benefits of taught strategies. Modeling by teacher of steps: statements or activities in the treatment descriptions that involve modeling from teacher in terms of demonstration of processes and/or steps the students are to follow.
62. Reinforcement	a. b.	No. Yes.	Probing-reinforcement: statements in the treatment description about intermittent or consistent use of probes, daily feedback, fading of prompts and cues, and/or overt administration of rewards and reinforcers.
63. Nonteacher instruction	a. b.	No. Yes.	Statements in the treatment description about homework, modeling from peers, parents providing instruction, and/or peers presenting or modeling instruction.

TABLE B.1 (continued)

CATEGORY	CODE OPTIONS		DEFINITION OF CODE
64.			
65. Directed response	a.	No.	Directed response/questioning: treatment description related to dialectic or Socratic teaching, the teacher directing students to ask questions, the teacher and student or students engaging in dialogue, and/or the teacher asks questions.
	a.	No.	
	b.	Yes.	
66. One-to-one instruction	a.	No.	Statements in the treatment description about activities related to independent practice, tutoring, instruction that is individually paced, and/or instruction that is individually tailored.
	b.	Yes.	
67. Technology	a.	No.	Statements in the treatment description about utilizing formal curriculum, newly developed pictorial representations, using specific material or computers, and/or using media to facilitate presentation and feedback.
	b.	Yes.	
68. Elaboration	a.	No.	Statements in the treatment description about additional information or explanation provided about concepts, procedures or steps, and/or redundant text or repetition within text.
	b.	Yes.	
69. Group instruction	a.	No.	Statements in the treatment description about instruction in a small group, and/or verbal interaction occurring in a small group with students and/or teacher.
	b.	Yes.	
70. Supplement	a.	No.	Supplement to teacher involvement besides peers: statements in the treatment description about homework, parent helps reinforce instruction. Using diagrams, pictures, or concrete materials to represent abstract information or concepts.
	b.	Yes.	
71. Strategy	a.	No.	Strategy cues: statements in the treatment description about reminders to use strategies or multisteps, the teacher verbalizing of steps or procedures to solve problems, use of think-aloud models, and/or teacher presenting the benefits of strategy use of procedures.
	b.	Yes.	
72. Adjusting difficulty levels	a.	No.	Changing difficulty levels of material or tasks to accommodate learning characteristics.
	b.	Yes.	
73. Peer tutoring	a.	No.	Using peer tutoring between two students, working as a tutor or tutee with a single partner.
	b.	Yes.	
74. Guided practice	a.	No.	Guiding students' independent practices (more learner-directed intervention).
	b.	Yes.	
75. Motivation-based intervention	a.	No.	Providing rewards or reinforcements, encouraging positive attitude, setting up supportive environment, using goal-setting methods, or managing disruptive or problematic behavior.
	b.	Yes.	

86

TABLE B.1 (continued)

	CATEGORY		CODE OPTIONS	DEFINITION OF CODE
76.	Mastery criterion	a.	No.	Using specific mastery criteria to specify a level of performance.
		b.	Yes.	
77.	Specific feedback	a.	No.	Providing various types feedback to students on performance such as completion ratio, growth lines, or graphed performance levels.
		b.	Yes.	
78.	External rewards	a.	Individual external rewards.	
		b.	Group external rewards.	
		c.	Positive teacher attitude.	
		d.	Student team learning.	
79.	Pull-out setting	a.	No.	Child "pulled-out" of classroom setting.
		b.	Yes.	
80.	Small group games	a.	No.	Use of small group games to teach number concepts.
		b.	Yes.	
81.	Feedback	a.	No.	Feedback given to students, with corrections.
		b.	Yes.	
82.	Fluency building	a.	No.	i.e., speed drills
		b.	Yes.	
83.	Parental involvement	a.	No.	
		b.	Yes.	
84.	Homework	a.	No.	
		b.	Yes.	
85.	Intervention goals	a.	School performance.	What intervention attempts to change/improve.
		b.	Psychological attributes.	
		c.	Both.	
		d.	Other.	
86.	Intervention description	a.	Fully detailed.	Level of description of intervention in study.
		b.	General.	
		c.	Descriptive label.	
		d.	No description.	

87

TABLE B.1 (continued)

	CATEGORY	CODE OPTIONS		DEFINITION OF CODE
87.	Practical utility	a.	No.	Can teachers implement the intervention easily in their classrooms?
		b.	Yes.	
88.		a.		
89.	Test administrator	b.	Ph.D. individual.	Description of test administrator.
		c.	Graduate student.	
		d.	Undergraduate.	
		e.	Combination of a & b.	
		f.	b & c.	
		g.	a & c.	
		h.	Teachers.	
		i.	Parents.	
		j.	Hired help.	
		k.	Not specified.	
90.	Intervention administration	a.	Individual.	Type of intervention administration.
		b.	Small group.	
		c.	Large group.	
		d.	Pairing.	
		e.	b & c.	
91.	Weeks of intervention	Number of weeks.		Number of weeks of treatment.
92.	Maintenance	a.	No.	Did the study mention maintenance?
		b.	Yes.	
93.	Transfer	a.	No.	Did the study describe a transfer effect?
		b.	Yes.	
94.	Material, books	a.	No.	Did the intervention materials include the use of books?
		b.	Yes.	
95.	Material, games	a.	No.	Did the intervention materials include the use of games?
		b.	Yes.	
96.	Material, flash cards	a.	No.	Did the intervention materials include the use of flash cards?
		b.	Yes.	
97.	Material, worksheets	a.	No.	Did the intervention materials include the use of worksheets?
		b.	Yes.	
98.	Material, hands-on activities	a.	No.	Did the intervention materials include the use of hands-on activities?
		b.	Yes.	
99.	Number of samples			Number of samples.

TABLE B.1 (continued)

CATEGORY		CODE OPTIONS	DEFINITION OF CODE
100. N			Sample size.
101. Number of treatment groups	a.	One.	Number of treatment groups in design.
	b.	Two.	
	c.	Three.	
	d.	More.	
102. Students in treatment	a.	One.	Number of students in the treatment group.
103. Number of control groups	b.	Two.	Number of control groups in design.
	c.	Three.	
	d.	Four.	
	e.	No control group.	
104. Students in control			Number of students in the control group.
105. Post-test N			Post-test total sample size.
106. Source of data	a.	Child.	Source of data collection.
	b.	Parent.	
	c.	Teacher.	
	d.	Other.	
	e.	Research assistant.	
	f.	c & e.	
107. Percentage of male students			Percentage of male students in study.
108. Percentage of female students			Percentage of female students in study.
109. Mean age			Students' mean age in months.
110. Standard deviation of age			Standard deviation of students' age.
111. Median age			Median age
112. Anglo			Percentage of Anglo students in study.
113. African-American			Percentage of African-American students in study.
114. Hispanic			Percentage of Hispanic students in study.
115. Asian			Percentage of Asian students in study.
116. Biracial			Percentage of biracial students in study.
117. Socioeconomic status	a.	Middle-class.	Socioeconomic status of student population in study.
	b.	Poverty.	
	c.	Other.	
	d.	Unknown.	
118. Middle-class			Percentage of middle-class students in study.

89

TABLE B.1 (continued)

CATEGORY		CODE OPTIONS	DEFINITION OF CODE
119. In poverty			Percentage of students in poverty.
120. Unknown			Percentage of students in unknown socioeconomic class.
121. Other classes			Percentage of students in other socioeconomic classes.
122. Population	a.	Preschool: 3-5 yrs.	Population of study.
	b.	Kindergarten: 5-6 yrs.	
	c.	First grade: 6-7 yrs.	
	d.	Head start children.	
123. Response rate			Response rate of study.
124. Attrition rate of treatment group			Percentage of mortality in treatment group.
125. Attrition rate of control group			Percentage of mortality in control group.
126. Method of data collection	a.	Games.	Method of data collection.
	b.	Pictures/test.	
	c.	Questionnaires.	
	d.	Puppets.	
	e.	Interview.	
	f.	Other.	
127. Measure	a.	Standardized.	Type of measure.
	b.	Interest/attitude.	
	c.	Both b & a.	
	d.	Specific skill test.	
	e.	a & d.	
128. Measure ID	a.	TEMA-2.	Measure identification.
	b.	PPVT.	
	c.	Woodcock Johnson.	
129. Counterbalancing	a.	No.	Did the study use counterbalancing measures?
	b.	Yes.	
130. Manipulation check	a.	No.	Were manipulation checks employed?
	b.	Yes.	
131. Trainer assignment	a.	Random.	Type of assignment for treatment administrators.
	b.	Nonrandom.	

90

TABLE B.1 (continued)

CATEGORY	CODE OPTIONS		DEFINITION OF CODE
132. Trainer	a. b. c. d.	Teacher. Researcher. Layperson. Other.	Person implementing intervention/treatment.
133. Trainer gender	a. b. c.	Male. Female. Unknown.	Gender of the trainer.
134. Trainer variation	a. b. c.	Different trainer. Same trainer. Cannot determine.	Did the intervention trainer vary during the treatment period?
135. Trainer training	a. b.	No. Yes.	Did the study provide trainer training?
136. Setting variation	a. b. c.	Different setting. Same setting. Cannot determine.	Did the intervention setting vary during treatment?
137. Classroom grouping	a. b. c.	Ability. Random. Not specified.	Grouping methods of control/treatment groups within a classroom.
138. Experimental design	a. b. c.	Experimental. Quasi-experimental. Correlation.	The experimental design of study.
139. Type of design	a. b. c.	Between. Within. Split-plot.	Design of experiment.

91

TABLE B.1 (continued)

CATEGORY	CODE OPTIONS	DEFINITION OF CODE
140. Assignment	a. Random selection. b. Random assignment of subjects. c. Random assignment of intact groups (class, school, or districts). d. Random assignment matched. e. Stratified groups. f. Equivalence at pretest. g. Related variables. h. Nonrandom assignment but matched groups. i. Nonrandom and non-matched groups.	Assignment of treatment/control groups.
141. Longitudinal/Cross-sectional	a. Longitudinal. b. Cross-sectional pre/post.	Longitudinal or cross-sectional study.
142. Quality of control	a. No treatment. b. Dummy treatment. c. Some type of training.	Type of control group.
143. Comparability of curriculum or material	a. No. b. Yes.	Were the control and the experimental groups comparable in terms of curriculum or material?
144. Comparability of instructors	a. No. b. Yes.	Were the control and the experimental groups comparable in terms of instructors?
145. Comparability of duration	a. No. b. Yes.	Were the control and the experimental groups comparable in terms of duration?
146. Pretest differences	a. Favors treatment. b. Favors control. c. Favors neither.	Difference in treatment and control groups' pretest stats.
147. Age differences	a. Favors treatment. b. Favors control. c. Favors neither.	Difference in age between treatment and control groups.
148. Treatment pre-test/post-test	a. Gain. b. Loss. c. No difference. d. Varied.	Change in pre-test to post-test of treatment group as compared to control group; measure 1.

TABLE B.1 (continued)

CATEGORY	CODE OPTIONS		DEFINITION OF CODE
149. Treatment pre-test/post-test 2	a.	Gain.	Change in pre-test to post-test of treatment group as compared to control group; measure 2.
	b.	Loss.	
	c.	No difference.	
	d.	Varied.	
150. Treatment pre-test/post-test 3	a.	Gain.	Change in pre-test to post-test of treatment group as compared to control group; measure 3.
	b.	Loss.	
	c.	No difference.	
	d.	Varied.	
151. Interval	a.	Immediately.	Interval between post tests and generalization tests.
	b.	After 1 week.	
	c.	After 2 weeks.	
	d.	After 3-4 weeks.	
	e.	After 5 weeks or more.	
152. Control DV task			Time spent on dependent variable task in the control group.
153. Type 1 error	a.	No control.	Did the study control for Type 1 error?
	b.	Control.	
154. Constant variables			What were the variables held constant in the study?
155. Items in final scale	a.	Standardized.	Number of items in final scale if not standardized test; measure 1.
	b.	(Enter number)	
156. Items in final scale, 2	a.	Standardized.	Number of items in final scale if not standardized test; measure 2.
	b.	(Enter number)	
157. Items in final scale, 3	a.	Standardized.	Number of items in final scale if not standardized test; measure 3.
	b.	(Enter number)	
158. Dimensions			Number of dimensions in study.
159. Required response, 1	a.	Selection.	Required response from scale 1.
	b.	Produce.	
	c.	Both.	
	d.	Not specified.	
160. Required response, 2	a.	Selection.	Required response from scale 2.
	b.	Produce.	
	c.	Both.	
	d.	Not specified.	

TABLE B.1 (continued)

CATEGORY	CODE OPTIONS		DEFINITION OF CODE
161. Required response, 3	a.	Selection.	Required response from scale 3.
	b.	Produce.	
	c.	Both.	
	d.	Not specified.	
162. Scale, measure 1	a.	Likert scale.	Type of scale in measure 1.
	b.	Rating scale.	
	c.	Dichotomous (yes/no).	
	d.	Other. Knowledge.	
	e.	Standardized.	
163. Scale, measure 2	a.	Likert scale.	Type of scale in measure 2.
	b.	Rating scale.	
	c.	Dichotomous (yes/no).	
	d.	Other. Knowledge.	
	e.	Standardized.	
164. Scale, measure 3	a.	Likert scale.	Type of scale in measure 3.
	b.	Rating scale.	
	c.	Dichotomous (yes/no).	
	d.	Other. Knowledge.	
	e.	Standardized.	
165. Test format	a.	Timed.	Timing format of test.
	b.	Untimed.	
166. Repeated measures	a.	No.	Are measures repeated?
	b.	Yes.	
167. Matching	a.	Matching occurred.	Did matching occur in the study?
	b.	No matching.	
168. Mean type, 1	a.	Arithmetic.	Type of reported mean; measure 1.
	b.	Median.	
	c.	Proportion.	
	d.	Other.	
	e.	a & d.	

TABLE B.1 (continued)

CATEGORY	CODE OPTIONS		DEFINITION OF CODE
169. Mean type, 2	a.	Arithmetic.	Type of reported mean; measure 2.
	b.	Median.	
	c.	Proportion.	
	d.	Other.	
	e.	a & d.	
170. Mean type, 3	a.	Arithmetic.	Type of reported mean; measure 3.
	b.	Median.	
	c.	Proportion.	
	d.	Other.	
	e.	a & d.	
171. Variance	a.	Standard deviation.	Type of variance reported.
	b.	Variance.	
	c.	Standard error.	
	d.	Proportion.	
	e.	Other.	
172. Descriptive stat	a.	No.	Reported descriptive statistics.
	b.	Yes.	
173. Significant stats	a.	ANOVA.	Reported significant test statistics.
	b.	Multiple regression.	
	c.	ANCOVA.	
	a.	Chi squares.	
	b.	T-test.	
	c.	t, F.	
	d.	Multiple comparisons.	
174. Effect size			Calculated effect size of study.

95

TABLE B.1 (continued)

CATEGORY	CODE OPTIONS		DEFINITION OF CODE
175. Reliability type, measure 1	a.	Cronbach alpha.	Type of reliability and estimate; measure 1.
	b.	KR 20/21.	
	c.	Hoyt's method.	
	d.	Test-retest.	
	e.	Split-half.	
	f.	Parallel tests.	
176. Reliability type, measure 2	a.	Cronbach alpha.	Type of reliability and estimate; measure 2.
	b.	KR 20/21.	
	c.	Hoyt's method.	
	d.	Test-retest.	
	e.	Split-half.	
	f.	Parallel tests.	
177. Reliability type, measure 3	a.	Cronbach alpha.	Type of reliability and estimate; measure 3.
	b.	KR 20/21.	
	c.	Hoyt's method.	
	d.	Test-retest.	
	e.	Split-half.	
	f.	Parallel tests.	
178. Reliability, measure 1			Reliability of scale; measure 1.
179. Reliability, measure 2			Reliability of scale; measure 2.
180. Reliability, measure 3			Reliability of scale; measure 3.
181. Interrater reliabilities	a.	No.	Were interrater reliabilities reported?
	b.	Yes.	
182. Convergent validity, 1			Convergent validity correlation; measure 1.
183. Convergent validity, 2			Convergent validity correlation; measure 2.
184. Convergent validity, 3			Convergent validity correlation; measure 3.
185. Convergent validity name, 1			Name of convergent validity; measure 1.
186. Convergent validity name, 2			Name of convergent validity; measure 2.
187. Convergent validity name, 3			Name of convergent validity; measure 3.
188. Discriminant validity, 1			Discriminant validity correlation; measure 1.
189. Discriminant validity, 2			Discriminant validity correlation; measure 2.
190. Discriminant validity, 3			Discriminant validity correlation; measure 3.
191. Discriminant validity name,			Name of discriminant validity; measure 1.

TABLE B.1 (continued)

CATEGORY	CODE OPTIONS	DEFINITION OF CODE
192. Discriminant validity name, 2		Name of discriminant validity; measure 2.
193. Discriminant validity name, 3		Name of discriminant validity; measure 3.
194. Author's view	a. Success b. Mixed. c. Failure. d. No conclusion.	Author's interpretation of study.

APPENDIX C

DESCRIPTION OF STUDIES USED IN THIS META-ANALYSIS

TABLE C.1

DESCRIPTION OF STUDIES USED IN THIS META-ANALYSIS

	AUTHOR	ID	YEAR	TITLE	JOURNAL
1.	Arnold, Fischer, Doctoroff, & Dobbs				
2.	Avesar	2	2002	Accelerating math development in Head Start classrooms.	Journal of Educational Psychology
3.	Ciancio, Rojas, McMahon, & Pasnak	150	Unpub.	The development of quantitative skills in preschool children	
4.	Clements	8	2001	Teaching oddity and insertion to Head Start children: an economical cognitive intervention.	Applied Developmental Psychology
5.	Clements & Sarama	32	1984	Training effects on the development and generalization of Piagetian logical operations and knowledge of number.	Journal of Educational Psychology
6.	DeLong & Sophian	74	Unpub.	Effects of a preschool mathematics curriculum: research on the NSF-funded Building Blocks project.	
7.	Field	139	unpub	Young children solve numerical ratio problems with an intervention based on intuitive knowledge	
8.	Fischer	49	1981	Can preschool children really learn to conserve?	Child Development
9.	Gelman	77	1990	A part-part-whole curriculum for teaching number in the kindergarten.	Journal for research in Mathematical Education
10.	Hong	50	1982	Accessing one-to-one correspondence: Still another paper about conservation.	British Journal of Psychology
11.	Kaul	14	1996	Effects of Mathematics Learning through Children's Literature on Math Achievement and Dispositional Outcomes	Early Childhood Research Quarterly
12.	Kim, Charmaraman, Klein, & Starkey	46	Unpub.	Starting children too early on number work: A mismatch of developmental and academic priorities.	
13.		15	Unpub.	Supporting the development of informal mathematical knowledge in young children.	
14.	Malabonga & Pasnak	67	1995	Cognitive gains for kindergarteners instructed in seriation and classification.	Child Study Journal
15.	Malofeeva, Day, Saco, Young, & Ciancio	70	Unpub.	Developing number sense through instruction with Head Start children	
16.	Malofeeva, Day, Saco, & Young	71	Unpub.	Effects of number sense training	

99

TABLE C.1 (continued)

	AUTHOR	ID	YEAR	TITLE	JOURNAL
17.	Malofeeva, Ciancio, & Day	72	Unpub.	Development of Emergent math and literacy skills	Psychological Studies
18.	Mohanty & Mishra	29	1994	Effects of training on the developmental and generalization of Piagetian number skills and logical operations.	
19.	Pasnak	78	1987	Acceleration of cognitive development of kindergarteners.	Psychology in the Schools
20.	Pasnak, Hansbarger, Dodson, Hart, & Blaha	135	1996	Differential results of instruction at the preoperational/concrete operational transition.	Psychology in the Schools
21.	Pasnak, McCutcheon, Holt, & Campbell	17	1991	Cognitive and achievement gains for kindergartners instructed in Piagetian operations.	Journal of Educational Research
22.	Solter & Mayer	53	1978	Broader transfer produced by guided discovery of number concepts with preschool children.	Journal of Educational Psychology
23.	Sophian	138	Unpub.	Mathematics for the Future: Developing a Head Start Curriculum to Support Mathematics Learning	
24.	Sophian & Madrid	25	2003	Young children's reasoning about many-to-one correspondences.	Child Development
25.	Starkey & Klein	82	2000	Fostering parental support for children's mathematical development: An intervention with Head Start families.	Early Education and Development
26.	Tzuriel, Kaniel, Kanner, & Haywood	28	1999	Effects of the "Bright Start" program in kindergarten on transfer and academic achievement.	Early Childhood Research Quarterly
27.	Van de Rijt & Van Luit	30	1998	Effectiveness of the additional early mathematics program for teaching children early mathematics.	Instructional Science
28.	Van Luit & Schopman	69	2000	Improving early numeracy of young children with special educational needs.	Remedial and Special Education
29.	White & Alexander	12	1986	Effects of training on four-year-olds' ability to solve geometric analogy problems	Cognition and Instruction
30.	Yawkey	47	Unpub.	Sociodramatic play effects on mathematical learning and adult rating of playfulness in five year olds.	

APPENDIX D

INTERVENTION STUDIES USING THREE TO SEVEN-YEAR-OLD CHILDREN

TABLE D.1

INTERVENTION STUDIES USING THREE TO SEVEN-YEAR-OLD CHILDREN

AUTHOR	N	AGE	DURATION	INTERVENTION	SKILLS	RESULTS	ES
Arnold, Fischer, Doctoroff, & Dobbs (2002)	103	M=53.18 mo	6 weeks 4 activities per day	An intervention promoted emergent math skills	Number ID number sense counting estimation computation writing numerals n-o correspondence comparison	Children in the experimental group scores significantly higher on TEMA than children in the control group	d=.439, p<.05
Avesar (1984)	24	M=49.45 mo	2 sessions	Instruction on number identification tasks by means of modeling in a game-like format	Number id counting	Children in the experimental group did not improve over the children in the control condition on number identification tasks	d=.194, p>.05
Ciancio, Rojas, McMahon, & Pasnak (2001)	71	M=48.70 mo	15 min a day from mid-January to mid-May	Children played games designed to teach them insertion and oddity principle	Oddity insertion	Children in the experimental condition who played the learning set games became superior at oddity and insertions with manipulable objects, and they also scored higher on numeracy and memory scales than children in the control group	d=.791, p<.05

TABLE D.1 (continued)

AUTHOR	N	AGE	DURATION	INTERVENTION	SKILLS	RESULTS	ES
Clements (1984)	45	$M=54.00$ mo	8 weeks	Experimental treatment was based on either the logical foundations model or a skills integration model	Number ID computation counting classification cardinality n-o correspondence comparison fraction conservation	Children in the experimental groups significantly outperformed the control group on number concepts and logical operations	$d=14.26$ $p<.01$
Clements, & Sarama (2002)	68	$M=49.90$ mo	As part of a school-based program implemented every day throughout the school year	Building Blocks, a research-based software and print curriculum	number sense counting computation writing numerals n-o correspondence comparison geometry and spatial sense measurement	Children in the experimental condition improved on geometry and number tests than children in the control group	$d=1.16$, $p<.01$
				protoquantitative ratio knowledge through spatial cues	conservation	performance on numerical ratio problems than children in the control condition	$d=.287$, $p>.05$
				Intervention such variables as identity, reversibility, and compensation explanations were manipulated in training number and length concepts		Only some children in the experimental conditions improved their performance on conservation. Identity was the most significant factor in conservation acquisition; reversibility was important as well, but compensation was of little value.	

103

TABLE D.1 (continued)

AUTHOR	N	AGE	DURATION	INTERVENTION	SKILLS	RESULTS	ES
Fischer (1990)	86.00	kindergartners	25 days of instruction	Part-part-whole curriculum stressed set-subset relationships	Number ID counting computation cardinality n-o correspondence comparison fraction problem solving place value	Children in the experimental condition participants developed a more mature concept of number, were more successful in solving word problems, and developed greater understanding of place value in the base-ten numeration system	d=.565, p=.01
Gelman (1982)	76 and 60	M=48.33 and M=48.25	1 training session	Training emphasized the specific cardinal value of sets as a function of the number of objects in a set	n-o correspondence conservation	Children in the experimental group passed the large set-size trails yet, they continued to experience difficulties with conservation tasks	d=.15, p>.05 d=.19, p>.05
Hong (1996)	57	4 years 2 months to 6 years 4 months	9 weeks	Experimental group was exposed to mathematics related storybook reading and discussion/play time related to readings	number sense counting computation classification n-o correspondence comparison conservation patterns special info	Children in the experimental group improved on classification , number combination, and shape tasks	d=1.37, p<.01
Kaul (1991)	60	4- and 5-year olds	Not known	Process-based number readiness intervention	number sense counting computation classification one-to-one correspondence patterns conservation writing numerals	Children in the experimental condition performed better than the control children on most measures of number readiness and number concept knowledge test	d=1.15, p<.01

104

TABLE D.1 (continued)

AUTHOR	N	AGE	DURATION	INTERVENTION	SKILLS	RESULTS	ES
Kim, Charmaraman, Klein, & Starkey	83 and 80	M=59 mo and M=59 mo	School year	Intervention was designed to promote informal math knowledge in the areas of numerical and spatial cognition	enumeration number sense sense geometry spatial patterns measurement	Some children in the experimental condition from a middle-income sample and one third of low-income children improved their math performance on the Child Math Assessment measure used in the study	d=.03, p>.05 d=.05, p>.05
Malabonga, & Pasnak (1995)	17	M=68 mo	15-20 min, 2-3 days a week for 3 months	Children in the experimental condition were taught to classify objects according to size, form, orientation, pattern, and type of object.	classification seriation	Children in the experimental group improved on seriation and classification than students receiving instruction on academic subjects	d=.629, p<.05
Malofeeva, Day, Saco, Young, & Ciancio	40	M=57.9 mo	20-25 min session twice a week for 3 weeks	Children participated in number sense instruction emphasizing counting and number id through hands-on games, discussions, and developmentally appropriate activities	Number sense counting number id	Children in the experimental condition improved on their understanding of number as assessed by their performance on counting and number id scales of Number Sense Test. Except for addition and subtraction, training effects did not generalize to other math skills	d=.647, p<.05
Malofeeva, Day, Saco, & Young	124	M=59.33 mo	8 weeks for 2-3 times a week	Children in the instructional condition learned to compare through activities or a combination of activities and reading	comparison	Children in the experimental conditions significantly improved in their performance as compared to the children's performance in the control condition. No difference between two instructional conditions were found.	d=.673, p<.01

TABLE D.1 (continued)

AUTHOR	N	AGE	DURATION	INTERVENTION	SKILLS	RESULTS	ES
Malofeeva, Ciancio, & Day	141	$M=53.44$ mo	3 three-week training sessions	Children were taught through hands-on games	number id counting n-o correspondence	Only some children in the experimental group improved their performance on skill-specific measures	$d=.207$, $p>.05$
Mohanty & Mishra (1994)	30	$M=54$ mo	4 weeks	Experimental treatment was based on either the logical foundations model or a skills integration model	conservation	Children in both experimental groups had a significantly higher performance on tests of number knowledge and logical operations. The number skills group significantly outperformed the logical operations group on the number knowledge test, and the logical operations group significantly outperformed the number skills group on the logical operations test.	$d=4.765$ $p<.01$
Pasnak (1987)	22	kindergartners	15 min per day 2-3 times a week for 4 months	"learning set instruction" through coaching, supportive correction, and feedback	classification seriation conservation	Children in the experimental condition made twice the gains of the control condition on the measure of GLR ability and matched their gains on reading and mathematics achievement.	$d=.812$, $p>.05$

TABLE D.1 (continued)

AUTHOR	N	AGE	DURATION	INTERVENTION	SKILLS	RESULTS	ES
Pasnak, Hansbarge, Dodson, Hart, & Blaha (1996)	45	$M=60$ mo	15 min per day 3 days a week 2 1/2 to 3 months	Instruction in the form of 40 short lessons including 160 manipulatives	classification seriation conservation	Children in the experimental condition did not score higher than children in the control condition on mathematics subscale of O-LSAT and Mathematics subscale of SESAT for both urban and suburban schools	$d=.154$, $p>.05$
Pasnak, McCutcheon, Holt, & Campbell (1991)	47	$M=72$ mo	15-20 min a day 3-4 days a week for 3 months	Unidimensional classification, unidimensional seriation, and number conservation were taught (a Piacceleration curriculum)	conservation	Children in the experimental condition significantly gained on the Otis-Lennon School Ability Test and on Stanford Early School Achievement Test subscales than children in the control condition	$d=.752$, $p<.05$
Solter & Mayer (1978)	24	$M=48.5$ mo	1 session	Children learned about one-to-one correspondence by matched recovery, expository, or observational methods of instruction	Counting computation measurement additive and multiplicative relations	Children in the experimental condition did not differ on short-term recall from children in the control condition	$d=-.05$, $p>.05$

TABLE D.1 (continued)

AUTHOR	N	AGE	DURATION	INTERVENTION	SKILLS	RESULTS	ES
Sophian	123	M=53.4mo	9 month curriculum	Measurement-based approach implemented through math curriculum emphasizing that a numerical example depends on a choice of a unit and that units can be combined to form high- or lower-order units	Enumeration measurement geometry	Children in the experimental condition had higher performance on DSC and a supplementary instrument than in literacy control or no-intervention control	d=.336, p>.05 d=.297, p>.05 (with literacy) d=.364, p>.05 (with no-intervention)
Sophian and Madrid (2003)	120	M=63 mo	6 training problems administered during one lesson	Intervention highlighted the iterative nature of many-to-one mappings (the procedure emphasized the mapping between individual large blocks and sets of several small blocks)	one-to-one correspondence analogy matching measurement	Children in the experimental group did not benefit from training when compared to children's performance in the control condition	d=.01, p>.05

TABLE D.1 (continued)

AUTHOR	N	AGE	DURATION	INTERVENTION	SKILLS	RESULTS	ES
Starkey & Klein (2000)	Study 1: 31 Study 2: 28	Study 1: 56 Study 2: 57	4 months (8 biweekly classes for parents and children and learning mathematics with parents at home using math kits)	African-American (Study 1) or Latino (Study 2) mothers learned a math curriculum with their pre-kindergarten children through offered classes. Used math kits at home in between sessions	Numerical reasoning Cardinality counting comparison Computation geometry	Children in the experimental condition had a higher math composite score than children in the control condition who received regular math instruction in their classroom	$d=.813$, $p<.05$ $d=.413$, $p>.05$
Tzuriel, Kaniel, Kanner, & Haywood	124	$M=57.15$ mo	10 months	The program used mediational teaching style, seven cognitive "small group" curriculum units, a cognitive-mediational management system, and a program for parent participation	Number concepts comparison classification sequences and patterns	Children in the experimental condition exposed to Bright Start program improved their performance on different cognitive tasks and showed more task-intrinsic motivation and metacognitive behavior than did those in the comparison group	$d=.565$, $p<.01$
Van de Rijt & Van Luit (1998)	106	$M=71$ mo	26 half an hour lessons 2 times a week	The 2 experimental groups were given a program that consisted of activities, embedded in real (daily) life themes, in which attention is paid to the different aspects of early mathematical competence.	Conservation counting	Children in guided and structured experimental groups significantly outperformed children in the control groups. No differences between two types of experimental groups were reported	$d=1.41$, $p<.01$

TABLE D.1 (continued)

AUTHOR	N	AGE	DURATION	INTERVENTION	SKILLS	RESULTS	ES
Van Luit & Schopman (2000)	124	$M=74.5$ mo	6 months 2 half-hour sessions per week	A curriculum for children with special educational needs emphasizing tally marks and a combination of guided and structured methods	Number ID counting computation classification ordinality n-o correspondence comparison	Children in the experimental condition scored higher on Early Numeracy Test but not on transfer test than children in the control condition	$d=.663$, $p<.01$
White, & Alexander (1986)	30	4-year olds	3 20-25 min instructional sessions	Training incorporated review, modeling, small and individual practices of how to solve geometric analogy problems	analogy comparison	Children in the experimental condition outperformed nontrained children on the geometric analogy task, and the effects of that training were maintained for 1 mo.	$d=2.68$, $p<.01$
Yawkey	96	$M=60$ mo	15 min daily for 7 months	Children in the experimental group learned math by means of sociodramatic play	not indicated	Children in the experimental group outperformed children in the control group on the Kindergarten Keys Mathematics Readiness Test	$d=.349$, $p>.05$

APPENDIX E: ADDITIONAL ANALYSIS

Between Group Comparisons Including the Outliers

The analyses yielded 29 studies that included 32 effect sizes (3 studies had two separate experiments in each one of them). The mean adjusted and unadjusted effect sizes for all studies for both fixed and random models are presented in Table 10. Effect sizes ranged from -.047 to 13.875. The mean effect size across all studies was .509 (*CI* from .415 to .604) for a fixed effect and .627 (*CI* from .416 to .837) for a random effect. Thus, the between-group comparisons were able to show benefit of training general over a control condition.

TABLE 10

MEANS OF EFFECT SIZES WITH OUTLIERS

Index	Standardized Difference (d) and 95% Confidence Interval			
	Point Estimate	SE	Lower Limit	Upper Limit
Cohen's d (unadjusted)				
Fixed	.803	.046	.712	.894
Random	1.117	.334	.462	1.772
Hedge's g (adjusted)				
Fixed	.509	.048	.415	.604
Random	.627	.107	.416	.837

NOTE: Number of outcomes=32. Number of individual cases=1935.

REFERENCES

Studies used in the meta-analysis have an asterisk next to them.

Anderson, A. (1997). Families and mathematics: A study of parent-child interactions. *Journal for Research in Mathematics Education, 28*(4), 484-511.

Arai, S. (1984). The development of counting. *Journal of Child Development, 20*, 13-19.

*Arnold, D. H., Fisher, P. H., Doctoroff, G. L., & Dobbs, J. (2002). Accelerating math development in Head Start classrooms. *Journal of Educational Psychology, 94*(4), 762-770.

*Avesar, C. (1984). The understanding of one to one correspondence by preschool children. *Dissertation Abstracts International, 46*(3-B), 975.

Baker, S., Gersten, R., & Lee, D. S. (2002). A synthesis of empirical research on teaching mathematics to low-achieving students. *Elementary School Journal, 103*(1), 51-73.

Baroody, A. J. (1987a). *Children's mathematical thinking: A developmental framework for preschool, primary, and special education teachers.* New York, NY, US: Teachers College Press.

Baroody, A. J. (1987b). The development of counting strategies for single-digit addition. *Journal for Research in Mathematics Education, 18*(2), 141-157.

Baroody, A. J. (1989). Kindergartners' mental addition with single-digit combinations. *Journal for Research in Mathematics Education, 20*(2), 159-172.

Baroody, A. J. (1992a). The development of kindergartners' mental-addition strategies. *Learning and Individual Differences, 4*(3), 215-235.

Baroody, A. J. (1992b). The development of preschoolers' counting skills and principles. In C. Meljac (Ed.), *Pathways to number: Children's developing numerical abilities* (pp. 99-126). Hillsdale, NJ, US: Lawrence Erlbaum Associates, Inc.

Baroody, A. J. (1999). Children's relational knowledge of addition and subtraction. *Cognition and Instruction, 17*(2), 137-175.

Baroody, A. J., & Coslick, R. T. (1998). *Fostering children's mathematical power: An investigative approach to K-8 mathematics instruction.* Mahwah, NJ, US: Lawrence Erlbaum Associates, Inc., Publishers.

Baroody, A. J., & White, M. S. (1983). The development of counting skills and number conservation. *Child Study Journal, 13*(2), 95-105.

Baroody, A. J., & Wilkins, J. L. (1999). The development of informal counting, number, and arithmetic skills and concepts. In J. V. Copley (Ed.), *Mathematics in the early years* (pp. 48-65). Washington, DC, US: National Association for the Education of Young Children.

Bertelli, R., Joanni, E., & Martlew, M. (1998). Relationship between children's counting ability and their ability to reason about number. *European Journal of Psychology of Education, 13*(3), 371-384.

Bobis, J., & Whitton, D. (1999). *Count Me In Too 1998 report*: University of Sydney.

Boisvert, M., Standing, L., & Moller, L. (1999). Successful part-whole perception in young children using multiple-choice tests. *Journal of Genetic Psychology, 160*(2), 167-180.

Brown, A. L., & Campione, J. C. (1994). Guided discovery in a community of learners. In K. McGilly (Ed.), *Classroom lessons: Integrating cognitive theory and classroom practice* (pp. 229-270). Cambridge, MA: MIT Press.

Bryant, D. M., Clifford, R. M., & Peisner, E. S. (1991). Best practices for beginners: Developmental appropriateness in kindergarten. *American Educational Research Journal, 28*(4), 783-803.

Bryant, D. M., Clifford, R. M., & Peisner, E. S. (1994). Best practices for beginners. *American Educational Research Journal, 28*, 783-803.

Burts, M. (1993). Quality of center child care and infant cognitive and language development. *Child Development, 67*, 606-620.

Bus, A. G., & van Ijzendoorn, M. H. (1999). Phonological awareness and early reading: A meta-analysis of experimental training studies. *Journal of Educational Psychology, 91*(3), 403-414.

Bus, A. G., van Ijzendoorn, M. H., & Pellegrini, A. D. (1995). Joint book reading makes for success in learning to read: A meta-analysis on intergenerational transmission of literacy. *Review of Educational Research, 65*(1), 1-21.

Butler, F. M., Miller, S. P., Lee, K. h., & Pierce, T. (2001). Teaching mathematics to students with mild-to-moderate mental retardation: A review of the literature. *Mental Retardation, 39*(1), 20-31.

Campbell, K. C., & Maertens, N. (1988). Informal teaching of beginning arithmetic concepts to nursery and kindergarten children. *Journal of Instructional Psychology, 15*(1), 17-24.

Carpenter, T. P., Fennema, E., Fuson, K., Hiebert, J., Human, P., Murray, H., et al. (1999). Learning basic number concepts and skills as problem solving. In E. Fennema & T. A. Romberg (Eds.), *(1999). Mathematics classrooms that promote understanding. Studies in mathematical thinking and learning series* (pp. 45-61). Mahwah, NJ: Lawrence Erlbaum Associates, Publishers.

Case, R. (1998). *A psychological model of number sense and its development*: the annual meeting of the American Educational Research Association, San Diego.

Case, R., & Griffin, S. (1990). Child cognitive development: The role of central conceptual structures in the development of scientific and social thought. In C.-A. Hauert (Ed.), *Developmental psychology: Cognitive, perceptual motor and neuropsychological perspectives. Advances in psychology* (Vol. 64, pp. 193-230). Oxford, England: North-Holland.

Chao, S. J., Stigler, J. W., & Woodward, J. A. (2000). The effects of physical materials on kindergartners' learning of number concepts. *Cognition and Instruction, 18*(3), 285-316.

Charlesworth, R., & Lind, K. K. (1990). *Math and Science for Young Children*. Albany, NY: Delmar.

*Ciancio, D., Rojas, A. C., McMahon, K., & Pasnak, R. (2001). Teaching oddity and insertion to Head Start children. An economical cognitive intervention. *Journal of Applied Developmental Psychology, 22*(6), 603-621.

Ciancio, D., Sadovsky, A., Malabonga, V., Trueblood, L., & Pasnak, R. (1999). Teaching classification and seriation to preschoolers. *Child Study Journal, 29*(3), 193-205.

Clements, D. H. (1984a). The development of counting and other early number knowledge: A review of research and psychological models. *Psychological Documents, 14*(2), 22.

*Clements, D. H. (1984b). Training effects on the development and generalization of Piagetian logical operations and knowledge of number. *Journal of Educational Psychology, 76*(5), 766-776.

*Clements, D. H., & Sarama, J. (2003). Effects of a preschool mathematics curriculum: Summary research on Building Blocks project.

Clements, D. H., Swaminathan, S., Hannibal, M. A. Z., & Sarama, J. (1999). Young children's concepts of shape. *Journal for Research in Mathematics Education, 30*(2), 192-212.

Cobb, P. (1994). Where is the mind? Constructivist and sociocultural perspectives on mathematical development. *Educational Researcher, 23*(7), 13-20.

Cohen, J. (1977). *Statistical power analysis for the behavioral sciences (rev. ed.)*. Hillsdale, NJ, England: Lawrence Erlbaum Associates, Inc.

Cook, T. D., Cooper, H., Cordray, D. S., Hartmann, H., Hedges, L. V., Light, R. J., et al. (1992). *Meta-analysis for explanation: A casebook.* New York, NY: Russell Sage Foundation.

Cortina, J. M., & Nouri, H. (2000). *Effect size for ANOVA design.* Thousand Oaks, CA: SAGE.

Davie, R., Butler, N., & Goldstein, H. (1972). *From birth to seven: The second report of the Child Development Study (1958 cohort).* Oxford, England: Longman.

*DeLong, C. M., & Sophian, C. (2003). Young children solve numerical ratio problems with an intervention based on intuitive knowledge.

Dixon, R. C., Carnine, D. W., Lee, D. S., Wallin, J., & Chard, D. (1998). *Review of High Quality Experimental Mathematics Research.* California State Board of Education.

Education, U. D. (1991). *AMERICA 2000: An education strategy.* Washington, DC: U.S. Department of Education.

Elbaum, B., Vaughn, S., Hughes, M. T., Moody, S. W., & Schumm, J. S. (2000). How reading outcomes of students with disabilities are related to instructional grouping formats: A meta-analytic review. In R. Gersten & E. P. Schiller (Eds.), *Contemporary special education research: Syntheses of the knowledge base on critical instructional issues. The LEA series on special education and disability* (pp. 105-135). Mahwah, NJ: Lawrence Erlbaum Associates, Publishers.

Fennema, E., Carpenter, T. P., & Lamon, S. J. (Eds.). (1991). *Integrating research on teaching and learning mathematics.* Albany, NY: State University of New York Press.

*Field, D. (1981). Can preschool children really learn to conserve? *Child Development, 52*(1), 326-334.

*Fischer, F. E. (1990). A part-part-whole curriculum for teaching number in the kindergarten. *Journal for Research in Mathematics Education, 21*(3), 207-215.

Fuchs, D., Fuchs, L. S., Mathes, P. G., & Lipsey, M. W. (2000). Reading differences between low-achieving students with and without learning disabilities: A meta-analysis. In R. Gersten & E. P. Schiller (Eds.), *Contemporary special education research: Syntheses of the knowledge base on critical instructional issues. The LEA series on special education and disability* (pp. 81-104). Mahwah, NJ: Lawrence Erlbaum Associates, Publishers.

Fuson, K. C. (1992). Research on learning and teaching addition and subtraction of whole numbers. In R. T. Putnam (Ed.), *Analysis of arithmetic for mathematics teaching* (pp. 53-187). Hillsdale, NJ, England: Lawrence Erlbaum Associates, Inc.

Fuson, K. C. (1995). Aspects and uses of counting: An AUC framework for considering research on counting to update the Gelman/Gallistel counting principles. *Cahiers de Psychologie Cognitive/Current Psychology of Cognition, 14*(6), 724-731.

Fuson, K. C., & Kwon, Y. (1992). Learning addition and subtraction: Effects of number words and other cultural tools. In C. Meljac (Ed.), *Pathways to number: Children's developing numerical abilities* (pp. 283-306). Hillsdale, NJ, England: Lawrence Erlbaum Associates, Inc.

Fuson, K. C., Lyons, B. G., Pergament, G. G., Hall, J. W., & et al. (1988). Effects of collection terms on class-inclusion and on number tasks. *Cognitive Psychology, 20*(1), 96-120.

Fuson, K. C., Secada, W. G., & Hall, J. W. (1983). Matching, counting, and conservation of numerical equivalence. *Child Development, 54*(1), 91-97.

Gallistel, C. R., & Gelman, R. (1990). The what and how of counting. *Cognition, 34*(2), 197-199.

Gallistel, C. R., & Gelman, R. (1992). Preverbal and verbal counting and computation. *Cognition, 44*(1-2), 43-74.

Geary, D. C. (1994). *Children's mathematical development: Research and practical applications.*

Geary, D. C. (1995). Reflections of evolution and culture in children's cognition: Implications for mathematical development and instruction. *American Psychologist, 50*(1), 24-37.

*Gelman, R. (1982). Accessing one-to-one correspondence: Still another paper about conservation. *British Journal of Psychology, 73*, 209-220.

Gelman, R., & Gallistel, C. R. (1978). *The child's understanding of number.* Oxford, England: Harvard U Press.

Gersten, R., & Baker, S. (2000). The professional knowledge base on instructional practices that support cognitive growth for English-language learners. In R. Gersten & E. P. Schiller (Eds.), *Contemporary special education research: Syntheses of the knowledge base on critical instructional issues. The LEA series on special education and disability* (pp. 31-79). Mahwah, NJ: Lawrence Erlbaum Associates, Publishers.

Gersten, R., & Chard, D. (1999). Number sense: Rethinking arithmetic instruction for students with mathematical disabilities. *Journal of Special Education, 33*(1), 18-28.

Gersten, R., Schiller, E. P., & Vaughn, S. (Eds.). (2000). *Contemporary special education research: Syntheses of the knowledge base on critical instructional issues.* Mahwah, NJ: Lawrence Erlbaum Associates, Publishers.

Gifford, S. (1995). Number in early childhood. *Early Child Development and Care, 109*, 95-119.

Ginsburg, H. P., & Baron, J. (1993). Young children's construction of mathematics. In R. J. Jensen (Ed.), *Research Ideas for the Classroom: Early Childhood Mathematics* (pp. 3-21). Reston, VA: National Council of Teachers of Mathematics.

Ginsburg, H. P., Bempechat, J., & Chung, Y. E. (1992). Parent influences on children's mathematics. In T. G. Sticht & M. J. Beeler (Eds.), *The intergenerational transfer of cognitive skills, Vol. 1: Programs, policy, and research issues; Vol. 2: Theory and research in cognitive science. Cognition and literacy* (pp. 91-121). Westport, CT: Ablex Publishing.

Ginsburg, H. P., & Russell, R. L. (1981). Social class and racial influences on early mathematical thinking. *Monographs of the Society for Research in Child Development, 46*(6), 69.

Glass, G. V., McGaw, B., & Smith, L. B. (1981). *Meta-analysis in Social Research.* Beverly Hills, CA: Sage.

Griffin, S., Case, R., & Capodilupo, A. (1995). Teaching for understanding: The importance of the central conceptual structures in the elementary mathematics curriculum. In e. al. (Ed.), *Teaching for transfer: Fostering generalization in learning* (pp. 123-151).

Griffin, S., Case, R., & Siegler, R. S. (1994). Rightstart: Providing the central conceptual prerequisites for first formal learning of arithmetic to students at risk for school failure. In K. McGilly (Ed.), *Classroom lessons: Integrating cognitive theory and classroom practice* (pp. 25-49). Cambridge, MA: The MIT Press.

Groen, G., & Resnick, L. B. (1977). Can preschool children invent addition algorithms? *Journal of Educational Psychology, 69*(6), 645-652.

Hart, C. H., Burts, D. C., & Charlesworth, R. (Eds.). (1997). *Integrated curriculum and developmentally appropriate practice: Birth to age eight.* Albany, NY, US: State University of New York Press.

Hiebert, J., & Carpenter, T. P. (1992). Learning and teaching with understanding. In D. A. Grouws (Ed.), *(1992). Handbook of research on mathematics teaching and learning: A project of the National Council of Teachers of Mathematics* (pp. 65-97). New York, NY, England: Macmillan Publishing Co, Inc.

Holland, V. M., & Palermo, D. S. (1975). On learning "less": Language and cognitive development. *Child Development, 46*(2), 437-443.

*Hong, H. (1996). Effects of mathematics learning through children's literature on math achievement and dispositional outcomes. *Early Childhood Research Quarterly, 11*, 477-494.

Hughes, M. (1981). Can preschool children add and subtract? *Educational Psychology, 1*(3), 207-219.

Hughes, M. (1983). Teaching arithmetic to preschool children. *Educational Review, 35*(2), 163-173.

Hughes, M. (1986). *Children and Number: Difficulties in Learning Mathematics.* Oxford: Basil Blackwell.

Hughes, M. (1996). Parents, teachers and schools. In B. Bernstein & J. Brannen (Eds.), *Children, research and policy: Essays for Barbara Tizard* (pp. 96-110). Philadelphia, PA: Taylor & Francis.

Hunter, J. E., & Schmidt, F. L. (1989). Meta-analysis: Facts and theories. In M. Smith & I. T. Robertson (Eds.), *(1989). Advances in selection and assessment* (pp. 203-216). Oxford, England: John Wiley & Sons.

Hunting, R. P., & Sharpley, C. F. (1988). Fraction knowledge in preschool children. *Journal for Research in Mathematics Education, 19*(2), 175-180.

Jennings, C. M., Jennings, J. E., Richey, J., & Dixon-Krauss, L. D. (1992). Increasing interest and achievement in mathematics through children's literature. *Early Childhood Research Quarterly, 7*(2), 263-276.

Jitendra, A. K., Griffin, C. C., McGoey, K., Gardill, M. C., Bhat, P., & Riley, T. (1998). Effects of mathematical word problem solving by students at risk or with mild disabilities. *Journal of Educational Research, 91*(6), 345-355.

Jordan, N. C., Huttenlocher, J., & Levine, S. C. (1992). Differential calculation abilities in young children from middle- and low-income families. *Developmental Psychology, 28*(4), 644-653.

Kamii, C., Lewis, B. A., & Kirkland, L. (2001). Manipulatives: When are they useful? *Journal of Mathematical Behavior, 20*(1), 21-31.

Kaul, V., Bhatnagar, R., & Tolani, S. (1991). *Starting children too early on number work: A mismatch of developmental and academic priorities.* India.

Kimchi, R. (1993). Basic-level categorization and part-whole perception in children. *Bulletin of the Psychonomic Society, 31*(1), 23-26.

Kotovsky, L., & Gentner, D. (1996). Comparison and categorization in the development of relational similarity. *Child Development, 67*(6), 2797-2822.

Kroesbergen, E. H., & van Luit, J. E. H. (2002). Teaching multiplication to low math performers: Guided versus structured instruction. *Instructional Science, 30*(5), 361-378.

Kroesbergen, E. H., & Van Luit, J. E. H. (2003). Mathematics interventions for children with special educational needs: A meta-analysis. *Remedial and Special Education, 24*(2), 97-114.

Leder, G. C. (1992). Mathematics and gender: Changing perspectives. In D. A. Grouws (Ed.), *Handbook of research on mathematics teaching and learning: A project of the National Council of Teachers of Mathematics* (pp. 597-622). New York, US: Macmillan Publishing Co, Inc.

Lee, D. S. (2000). *A meta-analysis of mathematics interventions reported for 1971--1998 on the mathematics achievement of students identified with learning disabilities and students identified as low achieving.* University of Oregon.

Levine, S. C., Jordan, N. C., & Huttenlocher, J. (1992). Development of calculation abilities in young children. *Journal of Experimental Child Psychology, 53*(1), 72-103.

Lipsey, M. W., & Wilson, D. B. (2000). *Practical meta-analysis.* Thousand Oaks,CA: SAGE.

Lonigan, C. J., Burgess, S. R., Anthony, J. L., & Barker, T. A. (1998). Development of phonological sensitivity in 2- to 5-year-old children. *Journal of Educational Psychology, 90*(2), 294-311.

Maccini, P., & Hughes, C. A. (1997). Mathematics interventions for adolescents with learning disabilities. *Learning Disabilities Research and Practice, 12*(3), 168-176.

Malofeeva, E. V., Ciancio, D., & Day, J. D. (unpub.). *Development of emergent math and literacy skills.* University of Notre Dame.

Malofeeva, E. V., & Day, J. D. (unpub.). *Development of comparison skills in Head Start children.* University of Notre Dame.

Malofeeva, E. V., Day, J. D., Saco, X., Young, L., & Ciancio, D. (unpub.). *Developing number sense through instruction with Head Start children.* University of Notre Dame.

Marquis, J. G., Horner, R. H., Carr, E. G., Turnbull, A. P., Thompson, M., Behrens, G. A., et al. (2000). A meta-analysis of positive behavior support. In R. Gersten & E. P. Schiller (Eds.), *Contemporary special education research: Syntheses of the knowledge base on critical instructional issues. The LEA series on special education and disability* (pp. 137-178). Mahwah, NJ: Lawrence Erlbaum Associates, Publishers.

Mastropieri, M. A., Bakken, J. P., & Scruggs, T. E. (1991). Mathematics instruction for individuals with mental retardation: A perspective and research synthesis. *Education and Training in Mental Retardation, 26*(2), 115-129.

Mathematics, N. C. o. T. o. (2000). *Principles and Standards for School Mathematics.* Reston, VA.: NCTM.

Miller, K. (1984). Child as a measurer of all things: Measurement procedures and the development of quantitative concepts. In C. Sophian (Ed.), *Origins of cognitive skills* (pp. 193-228). Hillsdale, NJ: Lawrence Erlbaum Associates.

*Mohanty, B., & Mishra, S. (1994). Effects of training on the development and generalization of Piagetian number skills and logical operations. *Psychological Studies, 39*(2-3), 73-83.

Munn, P., & Stephen, C. (1993). Children's understanding of number words. *British Journal of Educational Psychology, 63*(3), 521-527.

NCTM. (1994-95). *1994-95 Handbook: NCTM Goals, Leaders, and Positions.* Reston, VA.: NCTM.

NCTM. (2000). *Principles and Standards for School Mathematics.* Reston, VA.: NCTM.

NCTM. (2003). *Early childhood mathematics: Promoting good beginnings.* Reston, VA:NCTM.

Newman, R. S., & Berger, C. F. (1984). Children's numerical estimation: Flexibility in the use of counting. *Journal of Educational Psychology, 76*(1), 55-64.

Okolo, C. M., Cavalier, A. R., Ferretti, R. P., & MacArthur, C. A. (2000). Technology, literacy, and disabilities: A review of the research. In R. Gersten & E. P. Schiller (Eds.), *Contemporary special education research: Syntheses of the knowledge base on critical instructional issues. The LEA series on special education and disability* (pp. 179-250). Mahwah, NJ: Lawrence Erlbaum Associates, Publishers.

*Pasnak, R. (1987). Acceleration of cognitive development of kindergartners. *Psychology in the Schools, 24*(4), 358-363.

Pasnak, R., Brown, K., Kurkjian, M., Mattran, K., & et al. (1987). Cognitive gains through training on classification, seriation, and conservation. *Genetic, Social, and General Psychology Monographs, 113*(3), 293-321.

Pasnak, R., & Campbell, J. (1991). Short- and long-term follow-up results for cognitive interventions. *Bulletin of the Psychonomic Society, 29*(4), 307-310.

*Pasnak, R., Hansbarger, A., Dodson, S. L., Hart, J. B., & Blaha, J. (1996). Differential results of instruction at the preoperational/concrete operational transition. *Psychology in the Schools, 33*(1), 70-83.

Pasnak, R., Madden, S. E., Malabonga, V. A., Holt, R., & et al. (1996). Persistence of gains from instruction in classification, seriation, and conservation. *Journal of Educational Research, 90*(2), 87-92.

*Pasnak, R., McCutcheon, L., Holt, R. W., & Campbell, J. W. (1991). Cognitive and achievement gains for kindergartners instructed in Piagetian operations. *Journal of Educational Research, 85*(1), 5-13.

Pasnak, R., Whitten, J. C., Perry, P., Waiss, S., & et al. (1995). Achievement gains after instruction on classification and seriation. *Education and Training in Mental Retardation and Developmental Disabilities, 30*(2), 109-117.

Pellegrini, A. D., & Stanic, G. M. (1993). Locating children's mathematical competence: Application of the developmental niche. *Journal of Applied Developmental Psychology, 14*, 501-520.

Pepper, K. L., & Hunting, R. P. (1998). Preschoolers' counting and sharing. *Journal for Research in Mathematics Education, 29*(2), 164-183.

Phillips, D., & Stipek, D. (1993). Early formal schooling: Are we promoting achievement or anxiety? *Applied and Preventive Psychology, 2*(3), 141-150.

Piaget, J. (1965). *The child's conception of number*. Oxford, England: W. W. Norton and Co.

Pike, C. D., & Forrester, M. A. (1997). The influence of number-sense on children's ability to estimate measures. *Educational Psychology, 17*(4), 483-500.

Resnick, L. B. (1989). Developing mathematical knowledge. *American Psychologist, 44*(2), 162-169.

Sarama, J., & Clements, D. H. (2002). Building Blocks for young children's mathematical development. *Journal of Educational Computing Research, 27*(1-2), 93-110.

Saxe, G. B. (1991). *Culture and cognitive development: Studies in mathematical understanding*. Hillsdale, NJ, England: Lawrence Erlbaum Associates, Inc.

Secada, W. G., Fuson, K. C., & Hall, J. W. (1983). The transition from counting-all to counting-on in addition. *Journal for Research in Mathematics Education, 14*(1), 47-57.

Siegler, R. S. (1987). Strategy choices in subtraction. In J. A. Sloboda (Ed.), *(1987). Cognitive processes in mathematics. Keele cognition seminars, Vol. 1* (pp. 81-106). New York, NY: Clarendon Press/Oxford University Press.

*Solter, A., & Mayer, R. E. (1978). Broader transfer produced by guided discovery of number concepts with preschool children. *Journal of Educational Psychology, 70*(3), 363-371.

Sophian, C. (1987). Early developments in children's use of counting to solve quantitative problems. *Cognition and Instruction, 4*(2), 61-90.

Sophian, C. (1988). Limitations on preschool children's knowledge about counting: Using counting to compare two sets. *Developmental Psychology, 24*(5), 634-640.

Sophian, C. (1998). A developmental perspective in children's counting. In C. Donlan (Ed.), *The development of mathematical skills. Studies in developmental psychology* (pp. 27-46).

Sophian, C. (2002). Learning about what fits: Preschool children's reasoning about effects of object size. *Journal for Research in Mathematics Education, 33*(4), 290-302.

*Sophian, C. (2003). Mathematics for the Future: Developing a Head Start Curriculum to Support Mathematics Learning. University of Hawaii.

Sophian, C., Garyantes, D., & Chang, C. (1997). When three is less than two: Early developments in children's understanding of fractional quantities. *Developmental Psychology, 33*(5), 731-744.

*Sophian, C., & Madrid, S. (2003). Young Children's Reasoning about Many-to-One Correspondences. *Child Development, 74*(5), 1418-1432.

Sophian, C., & McCorgray, P. (1994). Part-whole knowledge and early arithmetic problem solving. *Cognition and Instruction, 12*(1), 3-33.

Southard, N. A., & May, D. C. (1996). The effects of pre-first-grade programs on student reading and mathematics achievement. *Psychology in the Schools, 33*(2), 132-142.

Squire, S., & Bryant, P. (2002). The influence of sharing on children's initial concept of division. *Journal of Experimental Child Psychology, 81*(1), 1-43.

*Starkey, P., & Klein, A. (2000). Fostering parental support for children's mathematical development: An intervention with Head Start families. *Early Education and Development, 11*(5), 659-681.

Stevenson, H. W., Chen, C., & Lee, S. (1993). Motivation and achievement of gifted children in East Asia and the United States. *Journal for the Education of the Gifted, 16*(3), 223-250.

Stipek, D. J., Feiler, R., Byler, P., Ryan, R., Milburn, S., & Salmon, J. M. (1998). Good beginnings: What difference does the program make in preparing young children for school? *Journal of Applied Developmental Psychology, 19*(1), 41-66.

Stipek, D. J., Feiler, R., Daniels, D., & Milburn, S. (1995). Effects of different instructional approaches on young children's achievement and motivation. *Child Development, 66*(1), 209-223.

Swanson, H. L. (1992). The relationship between metacognition and problem solving in gifted children. *Roeper Review, 15*(1), 43-48.

Swanson, H. L. (2000). What instruction works for students with learning disabilities? Summarizing the results from a meta-analysis of intervention studies. In R. Gersten & E. P. Schiller (Eds.), *Contemporary special education research: Syntheses of the knowledge base on critical instructional issues. The LEA series on special education and disability* (pp. 1-30). Mahwah, NJ: Lawrence Erlbaum Associates, Publishers.

Swanson, H. L., Carson, C., & Saches Lee, C. M. (1996). A selective synthesis of intervention research for students with learning disabilities. *School Psychology Review, 25*(3), 370-391.

Swanson, H. L., & Hoskyn, M. (1998). Experimental intervention research on students with learning disabilities: A meta-analysis of treatment outcomes. *Review of Educational Research, 68*(3), 277-321.

Swanson, H. L., Hoskyn, M., & Lee, C. (1999). *Interventions for students with learning disabilities. A meta-analysis of treatment outcomes.* New York, NY: Guilford Press.

Swanson, H. L., & Lussier, C. M. (2001). A selective synthesis of the experimental literature on dynamic assessment. *Review of Educational Research, 71*(2), 321-363.

Tyler, Z. E., Allen, J. A., & Pasnak, R. (1983). Instruction effects on size and distance judgments. *Perception and Psychophysics, 34*(2), 135-139.

*Tzuriel, D., Kaniel, S., Kanner, E., & Haywood, H. C. (1999). Effects of the "Bright Start" program in kindergarten on transfer and academic achievement. *Early Childhood Research Quarterly, 14*(1), 111-141.

*Van de Rijt, B. A. M., & Van Luit, J. E. (1998). Effectiveness of the Additional Early Mathematics program for teaching children early mathematics. *Instructional Science, 26*(5), 337-358.

*Van Luit, J. E. H., & Schopman, E. A. M. (2000). Improving early numeracy of young children with special educational needs. *Remedial and Special Education, 21*(1), 27-40.

VanDevender, E. M. (1986). Fingers are good for early learning. *Journal of Instructional Psychology, 13*(4), 182-187.

Vygotsky, L. S. (Ed.). (1978). *Mind in society: The development of higher psychological processes.* Cambridge: Harvard University Press.

Wachter, K. W., & Straf, M. L. (Eds.). (1990). *The future of meta-analysis.* New York, NY: Russell Sage Foundation.

*White, C. S., & Alexander, P. A. (1986). Effects of training on four-year-olds' ability to solve geometric analogy problems. *Cognition and Instruction, 3*(3), 261-268.

Wiegel, H. G. (1998). Kindergarten students' organization of counting in joint counting tasks and the emergence of cooperation. *Journal for Research in Mathematics Education, 29*(2), 202-224.

Williams, C. K., & Kamii, C. (1986). How do children learn by handling objects? *Young Children, 42*(1), 23-26.

Williams, R. C. (1980). Training effects on number comparison and class inclusion skills of kindergarten children. *Dissertation Abstracts International, 41*(3-A), 931.

Xin, Y. P., & Jitendra, A. K. (1999). The effects of instruction in solving mathematical word problems for students with learning problems: A meta-analysis. *Journal of Special Education, 32*(4), 207-225.

*Yawkey, T. (not known). *Sociodramatic play effects on mathematical learning and adult ratings of playfulness in five-year olds.* Pennsylvania State University, University Park.

Young Loveridge, J. M. (1987). Learning mathematics. *British Journal of Developmental Psychology, 5*(2), 155-167.

Young Loveridge, J. M. (1989). The development of children's number concepts: The first year of school. *New Zealand Journal of Educational Studies, 24*(1), 47-64.

Printed in Great Britain by
Amazon.co.uk, Ltd.,
Marston Gate.